ALGAL BIOFUEL: RECENT ADVANCES

ALGAL BIOFUEL: RECENT ADVANCES

Dr. Abida Taskeen

Delve
PUBLISHING

www.delvepublishing.com

Algal Biofuel: Recent Advances

Dr. Abida Taskeen

Delve Publishing
2010 Winston Park Drive,
2nd Floor
Oakville, ON L6H 5R7
Canada
www.delvepublishing.com
Tel: 001-289-291-7705
001-905-616-2116
Fax: 001-289-291-7601
Email: orders@arclereducation.com

ISBN: 978-1-77361-171-6 (Hardcover)

Delve Publishing publishes wide variety of books and eBooks. For more information about Delve Publishing and its products, visit our website at www.delvepublishing.com.

ABOUT THE AUTHOR

Dr Abida Taskeen has a PhD degree in Chemistry and is working as the Head of Quality Enhancement Cell at Qarshi University, Lahore, Pakistan. She has expertise in field of Education, Social Services, Environmental Health and Science and Technology. She played a tremendous role in the establishment of the Quality Enhancement Cell and self assessment process at Qarshi University. She is also a recourse person at Directorate of Staff Development (Punjab Government). She is an author of many research articles and books.

TABLE OF CONTENTS

LIST OF FIGURES

LIST OF TABLES

PREFACE

This book focuses on the importance of Algae biofuels. It highlights the various tools and technique that are used in the process of extraction of the biofuels from algae. The book is divided into eight chapters which deal with the Algal biofuel source, extraction process, and their uses in the market. Algae have been used commercially for decades, but not for the production of biofuels. Several factors are driving the search for renewed fuel. The consortium of oil depletion, desired dependable energy, the need for carbon neutral renewable fuels can be produced without comprising on the environmental health.

The first chapter of the book provides a comprehensive introduction of the algal biofuels. It also introduces us to the tools and protocols being used for the sustainable development of biofuelsderived from algae. It further highlights the concept of life cycle assessment for algal biofuels and presents different studies approaches in regards to the production of biofuels from algae.

The next chapter discusses various pathways that can be adapted for the production of biofuels. It explains all the steps that are necessary to process algae into biofuels. It also introduces and compares various cultivation systems that can be employed to process algal biomass in to fuels followed by the extraction and harvesting techniques for the collection of bio-fuels.

Third chapter in this book talks about the resources inputs (such as water, soil, energy, nutrients) that are requiredfor the production of algal biofuels. algal biofuels. In addition to that, it provides suggestions on how to choose the best algae to maximize the production of bio-fuel. It also presents the concepts of algal farms and artificial rainforests. The ideas are well supported and are based on the recent studies for the production of biofuel at commercial level. The production of algae bio-fuels is technically feasible in a number of ways; and economically and environmentally seems to become a viable substitute for petroleum fuels.

The fourth chapter of this book provides a comprehensive explanation of the various environmental impacts of algal bio-fuels including but not limited to the water and air quality, aquatic and terrestrial biodiversity. The concept of social acceptability of genetically engineered organisms and their impacts environment and its inhabitants have also been discussed.

Chapter five of this bookdemonstratesalgae oils as fuels. The chapter provides a mechanistic approach towards the synthesis of bio-fuels starting from raw material. It provide a comprehensive guide towards the biochemistry for lipid synthesis, substrates requirements for micro algae growth, micro-algal cultivation, followed by trans-esterification techniques towards the preparation of bio-fuels. At the end of this chapter algal fuel properties were also discussed.

Sixth chapter of this book focuses on algal biomass harvesting. It starts with stability and separability characteristics of microalgae. These two characteristics greatly affect the decision to choose on a particular harvesting technology and its performance. Various harvesting methods in use have been discussed in detail. The last part of this chapter explains the importance of bioremediation as the process does not involve the use of any chemical and allows the wastes to be recycled into useful product.

Chapter seven of this book deals will economic production issues related to the bio-fuels. It presents various methodologies for the cost analysis of algal biofuel production.

The last chapter of this talks about the next generation of bioalgal fuels for example green fuel from green algae. It also presents some data and information about the companies that are making biofuels from algae.

Hence, the book highlights methodological challenges, recent advances and opportunities towards the production of algal biofuels. The book would be of particular interest to the graduate and the post graduate students and researchers of Applied Biology, Biotechnology, Microbiology, Biochemistry and Chemical Engineering, working on algae biofuels.

Author

CHAPTER 1

INTRODUCTION OF ALGAL BIOFUEL

A Biofuel alga is a diverse fossil fuel liquid based on algae. There are many well-known sources of bio-fuels such as algae, sugar cane and corn etc.

Algal oil/bio fuel or simply algae fuel is alternative oil which uses algae. It's an alternative to fossil fuels (liquid) and other biofuels sources like sugarcane and corn. Alga biofuels are based on energy rich oils. Many Government agencies and companies are funding researches/efforts to reduce operating and capital cost so as to make algae fuel production economically feasible. Most attractive characteristic of the algae fuels is that they can be grown with very less impact on the resources of fresh water. Algae biofuels have relatively no harm if spilled and can be produced from waste/saline water and are biodegradable.

Algal fuels also release CO_2 like other fossil fuels but algal fuel (like other biofuels) only release CO_2 removed recently through photosynthesis during algae growth. An interest has been developed in farming the algae (algaculture) regarding food crises and energy crises. For this purpose land which is unsuitable for agriculture is used.

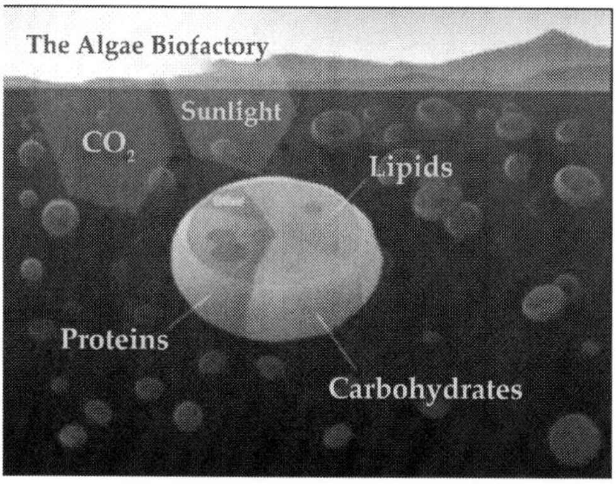

Figure 1.1: (a) Algae Bio-factory.

Being the rich source of fatty acids, algal oil is used in food products as supplement, particularly in milk because it has heavy contents of mono and poli-saturated fats. Beside this, the food supplement from agal oil is compatible with other diets e.g. vegetarian. In Docosahexaenoic acid (DHA), the contents are almost equivalent ot the fish oil based on salmon.

Algae Species

- *Chlorella*
- *Botryococcus braunii*
- *Dunaliella tertiolecta*
- *Pleurochrysis carterae*
- *Gracilaria*
- *Sargassum*,

The amount of biofuel/oil produced by the algae may vary with the specie.

ALGAL BIOFUEL

Algae biomass can be processed to create a variety of biofuels including ethanol or methanol.

According to the principle of association of biomass algae, the energy generated by the algae can reach the equivalent of oil in 2018. Figure 1.1 (b) represents the top ten countries for the production of biofuels and their production capacity in 2016-2017.

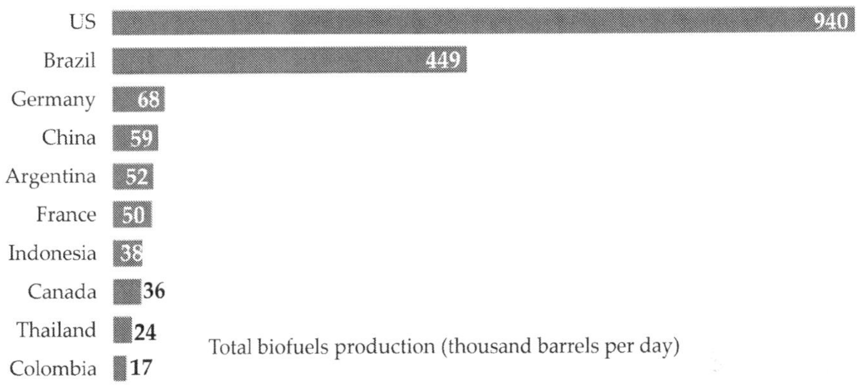

Figure 1.1: (b) Biofuel production capacities of top ten biofuel producing countries.

Petroleum and Diesel Based fuels

Oil-based fuel oil has been the main transport fuel on the planet for many years. Until 1960, the national production of algae bio-fuels met the huge mass of demand for the nation, but with the increasing population demand of the production to grow. They need to reduce their dependence on oil and need to look for alternative fuelsources. Biofuel algae and cyanobacteria have been considered as a promising alternative fuel to recover fuel production.

Macroalgae and microalgae grow by converting solar energy in chemical energy viaphotosynthesis. And some of algae species may grow even in heterotrophic conditions. In such situations, an exogenous source of carbon is expected to contribute to the production of fuels.

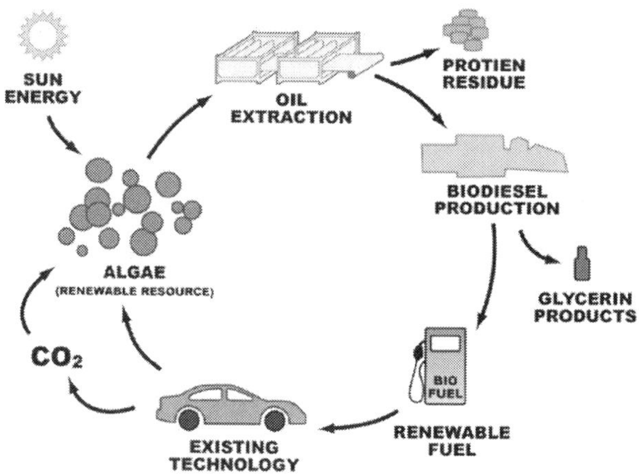

Figure 1.2: productions of Algal biofuels.

SUSTAINABLE DEVELOPMENT OF BIOFUELS:

Sustainable biofuels are produced in a sustainable manner as liquid fuels derived from plant material enter the market driven by many factors such as:

- peaks in oil prices and
- the need to increase energy security

However, many of these biofuels are criticized for their negative impact on natural environment, food security, and land.Sustainable development is a development that meets the needs of the present without compromising the ability of future generations to meet their needs.

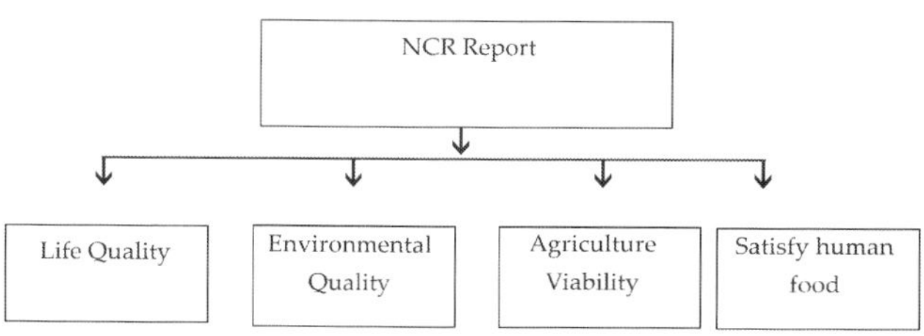

According to NRC report (recent) highlighted key social sustainability objectives for the agriculture. These goals are:

- The superiority of life improvement for farmers, farmers and society as a whole.

- Improved environmental superiority
- Maintaining the potential for profitable agriculture
- Satisfy human needs of food, feed and biofuels etc.

The above sustainabilityaspects are related. Some of these are synergistic and some can lead to exchange between goals. An example of synergycouldbeanimprovedtechnologyintheproductionofcyanobacteria and algae transformation of biomass fuels. These improvements should improve fuel efficiency, increase resource efficiency, energy security and reduce production costs and, consequently, contribute to the needs of transport fuel and improve environmental, economic and environmental sustainability. Organization of pollutants, an example of compensation, could be the, to maintain environmental quality and to minimize the impact on human health.

Component of Biofuel Development

Similarly as with biofuels, biofuels could give chances to enhance vitality security, lessen ozone depleting substance discharges, enhance asset base and keep up ecological quality. This is an impression of potential manageability issues that may emerge and anticipate or alleviate the advancement of the business. He examined the manageability of natural biofuels and different assessment criteria created in the course of the most recent ten years. A large number of the maintainability criteria apply to algae biofuels.

Security of Energy based on Available Resources

Depending on the availability of resources like water and land for the production of algae biofuels, the productivity of algae companies, and the ability to integrate the various components of biofuels. Constraints limit the amount of algae biofuels that may occur, but technological advances, it is possible to increase fuel oil the productivity and efficiency of algae materials.

Economics

The cost of production is an important sustainable development of all non-petroleum alternative fuels, including applied algae biofuels. While government subsidies and policies can facilitate and accelerate penetration in the biofuels market, biofuels must be economically viable without subsidies. Brazilian ethanol received heavy subsidies when the National Alcohol Braille Program started, but government subsidies were gradually eliminated in the nineties.

Environmental Effects and Resource Use

Earth, nutrients and water are needed for growing algae. US surface has been declining in recent decades, and the water level in some aquifers used for irrigation is declining. The lack of nutrients in agricultural crops in surface waters and their effects on the environment have increased.

Some concerns about the availability and use of resources and environment could be reduced by the development of biofuels from algae. This is due the reason that production of cyanobacteria and algae does not require high quality territorial resources such as sugar cane, maize for Production of ethanol and soya or other source of biodiesel. Cyanobacteria and algae can be grown in rich drainage or salt water. In addition, the enrichment of cyanobacteria and algae with CO_2plussomeother nutrients contribute to increase production from larger scale biosphere sources of photossintéticas algae emission, such as power plants using fossil fuels to integrate CO_2 emissions from the plant with the cultivation system of algae. Another clue is to identify sources of proximity to algal biomass waste structures such as urban waste water. Nutrients in reflux can be used by plant systems and are part of the regeneration treatment of nutrients. Therefore, an important concern is to find out potential sites for algae proximity to a CO_2 source, such as a nondrinking water source, fossil fuel power plants and such as salty and waste water.

Social Well Being

In spite of the fact that it is improbable that green growth and cyanobacteria to create biomass to seek top notch arrive, it might have social worries about soil utilize are considered in the improvement of green growth biofuels. For instance, the position of green growth and cyanobacteria in biomass creation in the US forsake could be seen as a decent utilization of low-esteem arrive by a few, however as an interruption into other unkempt terrains. Furthermore, the utilization of hereditarily altered living beings underway frameworks could impact social agreeableness. This report outlines how the utilization of assets and natural impacts of vast scale creation of green growth biofuels could influence the social worthiness of algal biofuels.

ASSESSMENT OF ALGAL BIOFUELS FOR SUSTAINABLE DEVELOPMENT: TOOLS AND PROTOCOLS

Combustion of algae focuses on different concepts i.e. the system analysis framework, sustainability indicators, stroke analyzes and future scenarios. Measurements or modeling results repeat the observation of indicators is used to represent or serve as a reference to impact impacts. LCA and analytical methodology are assessing the future use of resources and environmental impact. System analysis is an additional conceptual approach to assess the impact of biofuels on algae.

Framework for Systems Analysis

A research points out the systems approaches to emphasize the impact assessment and development of strategies to integrate all biological and physical processes, reviews, restrictions on socio-economic and engineering units worldwide. Different models are used to define the result or theeffective resultfor a given input of the common engagement system. The application of the methodology meant "clarifying the purpose in the solution, developing a complete description, formulating alternative solutions, and deploying its consequences." The framework for analysis is particularly applicable to algal biofuel systems. Of all current renewable energy, biofuels which are derived from plant and algae resources of alternatives for renewable energy, represent one of the challenges of integrating more complex systems.

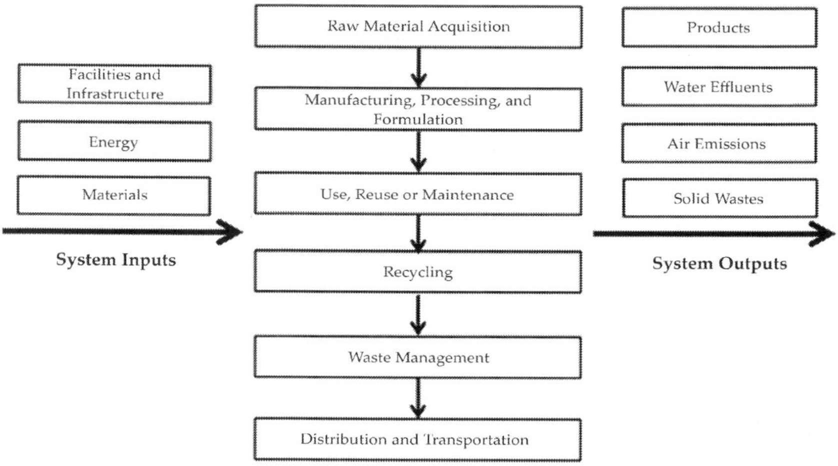

Figure 1.3: Schematic chart of the creation framework (counting framework information sources and yields).

Moreover, a considerable lot of these innovations are in a few phases of improvement, from a fascinating chance to pilot-scale exhibition. Notwithstanding the different structures, this many-sided quality includes that these advances can be incorporated to outline and assemble progressed biofuel frameworks. This assorted variety in blending advances and the conceivable joining administration is a development motor as at present found in the different business ways to deal with the improvement of green growth biofuels. In the meantime, this assorted variety makes issues for recording basic material, vitality, and stream quickly expected to assess execution. Understanding by receiving a structure for surveying elective outline frameworks for the generation of fluid fuel from green growth requires the selection of an option scoring plan. This framework outlines the free idea of the individual segments of the inventory network and the sources of info and yields of the framework. Understanding created by this portrayal is basic for an extensive variety of supportability apparatuses, for example, LCA, displaying designing procedures, and money saving advantage examination.

Indicator

Markers of Sustainability Indicators Define Sustainability Biofuels Status Accounting System or Progress Towards Sustainability Objectives. A few scientists and organizations to recognize the meanings of the one utilized as a part of this report is demonstrative "to guarantee that it is not the slightest bit characteristic of some great natural objectives, for example, ecological honesty, biological system wellbeing and supportable assets." The sign of a maintainable improvement of algal biofuels is circuitous, the mix of asset usage measurements, other ecological effects, social acknowledgment (all considered in this report) and vitality and vitality security (not considered in this report).

Since supportability incorporates ecological, financial and social measurements (notwithstanding vitality security and vitality, which can be ordered independently), markers are normally incorporated into these classifications. The classifications of asset needs markers that have been examined for biofuels incorporate aggregate utilization and water utilization, supplement utilize, add up to arrive utilize and net vitality recuperation and pointers classifications. The report features the manageability of the more extensive condition and along these lines introduces the classifications of marker pointers of natural perspectives that are applicable to algal biofuels. The report likewise underlines the economical utilization of assets that decides the legitimacy of the biofuel framework. This report introduces and breaks down pointers on the interface amongst natural and monetary supportability. Particular maintainability markers identified with different parts of the economy (eg worldwide exchange, benefit, work) are outside the span of this examination, albeit plainly the impact and impact of ecological manageability pointers. Social supportability markers of biofuels, frequently not considered or, but rather could create potential pointers for algal biofuels. Be that as it may, the target of this examination is manageability and markers identified with ecological effects and the requirements of common assets of nature.

Since manageable improvement is not advance towards maintainability objectives, it is critical to comprehend the nuts and bolts of supportability pointers. Furthermore, assigning some portion of the natural and social generation of green growth biofuels requires a comprehension of references and benchmarks. A fitting meaning of a standard ought to be built up without the creation of kelp biofuels. On a basic level, the standard consolidates dynamic soil utilize and natural changes related with the district, however by and by it is frequently more straightforward and more secure to consider the predominant conditions previously creating biofuels.

The utilization of specific units can impact the translation of supportability pointers. Units can incorporate the required volume or mass of assets; Concentrations of discharges, chemicals or ecological burdens; And the wealth of life forms are environment ranges, units can be created surface denominators of vitality or fuel volume. Picking a denominator, for example, the surface or volume of fuel, can make it less demanding to look at the utilization of soil or option fuel, however may likewise include vulnerability related with a pointer. For instance, the surface may incorporate framework or foundation zone, and a cushion. Counting time factor as marker permits considering the length of ecological impacts. Coefficients joining co-components of co-era can lead chiefs to decide the measure of an impact ought to be formally ascribed to the item.

Supportability and maintainable improvement grasp numerous objectives and objectives that influence dynamic human esteems. Advance towards supportability cannot be surveyed less targets and particular objectives are characterized and measurements are chosen for maintainability issues. Numerous universal associations are creating manageability markers for biofuels. These incorporate the Roundtable on Sustainable Biofuels (RSB), the Global Bioenergy Partnership (GBEP) bolstered by the G-8. There are different associations made to advance manageable biofuels, for example, feasible biomass consultancy in the United States. A worldwide confirmation framework for manageability and carbon biomass and bioenergy has been executed the world over by 750 players from 45 nations. A few associations prescribe an arrangement of supportability markers. For instance, RSB suggests more than 200 manageability markers for biofuels and measures. A test is to interface the biofuel supportability pointers expected to assess specific issues and make bland records in fact and monetarily attainable. Numerous endeavors to create nonspecific maintainability pointers for biofuels concentrated on cultivable biofuels, corn ethanol, cellulose biofuels and agrarian biodiesel. Subsequently, some prescribed markers may not be significant to algal biofuels, and some critical potential pointers may not show up in the already distributed records.

Affluenza et al. (2007) suggested that the accomplishment of definite markers in any circumstance. What more often than not prompts an appraisal of manageability is a choice or other reason in conjunction with supportability objectives. Maintainability objectives can incorporate ideas, for example, asset proficiency, water quality upkeep, biodiversity preservation and waste diminishment. Pointers ought to be mirrored the objectives.

An evaluation of maintainability for the creation of biofuels from green growth may incorporate contrasting algal biofuels and basic business situations (for instance, the utilization of fuel oil construct for the most part in light of the present premise) different sources (for instance) happen in Absence of biofuel generation or different locales for establishment. These examinations can lead evaluators to benefit various diverse maintainability pointers and can demonstrate how distinctive strategies and units of estimation or displaying.

LCA for Algal Biofuels

LCA is a set of databases, methods and tools which describe the impact of a service life cycleor product on environment. The LCA is defined as "a set of methodological procedures and compilation of energy and environmental input and output materials and impacts directly related to the performance of a product or service system during its lifecycle. "The life cycle in the context of biofuel production algae refers to a chain of activities that includes the extraction of crude resources, production of materials, manufacture, transportation, use and disposal".

Life cycle assessment takes into account an environmental impact or a resource. However its not possible to analysis all process in a life-cycle.

The LCA approach involves numerical modeling of material streams in chains. The concept is to assign a target product through a variety of processes or activities and use input/output tables to estimate material/ unit of product. Thisapproach has the advantage of having sector data quantifying the relation between the final products and energy source. However, due to the nascent nature of algae biofuel production, the LCA approach will focus on an LCA approach that uses specific nutritional information on each of the components and process resources. Focus Focus approach is a bottom-up approach built up of a complete estimation chain to examine individual components. Conversely, the LCA input-output approach is a type of top-down approach that uses a global model.

With the table of input and output, it is divided into sectors to describe economic transactions. Uncertainty in stroke models, the process method undergoes low-life variations in defining the outline of the system when the chain data portion is unavailable, while the LCA interrelation focus is an error associated with the aggregate product sectors. The life cycle assessment is a set of methods (hybrid) that combine the expected process.

In Life Cycle Assessment, another component is impact assessment. Impact assessment in terms of environmental impact, is interpretation flows life cycle in terms of environmental impact. An important impetus for impact assessment is the mapping of multiple impacts flows (e.g. resource availability, climate changeand human toxicity) including development. There should be decision-making tools reported to navigate these multiple impacts. Many LCAs for other experiences for the renewable fuels CNRC biofuels report: potential environmental and economic effects of biofuels policy in United States.

LCA can provide important support for information on algal biofuels. In case of algal biofuels have potentially less greenhouse gases and fuel cannot generate new negative impact on emissions. Depending on the type of raw material, estimates gas emissions of greenhouse in the life cycle of other biofuels cover a wide range; Different practices used for the cultivation of raw materials and, if there have been changes in land use. Therefore, algal biofuels should be examined with LCA and other approaches. In addition, extension systems contribute significantly to nutrients and water and emit more polluting substances. Flows linked with the standard materials listed above for the algae biofuel industry can be categorized by life cycle assessment.

There are a lot of challenges in using life cycle assessment to assess the algae biofuels' sustainability. The production of algae biofuels is developing and historical data is not available. In addition, advances which have been made technologically and the impact of the balance on future material flow, but are difficult to predict.

The LCA to survey the natural factors that are creating social LCA to think about the social effect of items, procedures or business and recognize potential zones of change. Social ECV is at a beginning time of improvement and has not yet achieved an agreement on affect classifications to incorporate and measure.

ALGAL BIOFUELS DOMESTIC DEVELOPMENT: HISTORY

Since mid of twentieth century, different advantages as feedstock of algae have been known. Although a commercially viable and scalable system at commercial scale has not developed yet but it has been observed that previous studies have explored approaches for technologies reported today.Algae has been used as a mean to produce energy in 1996 Several researches reported the production of natural gas through anaerobic digestion using Carbohydrate portion of algae.A researcher in 1978 reported that methane gas can be produced from algae sell with competitive rates regarding fossil fuels.

In 1940s; it was revealed that rich lipid contents can be obtained from different species of algae under different conditions which was further supported by studies in 1950s and 1960s; which revealed that key nutrients (e.g. silicon or nitrogen) starvation may cause this phenomenon. During early 1970s, when fuel prices raised up this idea gained attention as source of energy. This concept of use of lipid stores became the key factor for the DOE's Aquatic Species Program. This is the program which reflects most inclusive studies/efforts for production of fuels from algae. The program (1978-1996), supported research at DOE's NREL (National Renewable Energy Laboratory) which was primarily called as Solar Energy Research Institute. Under this program for aquatic species, approximately 25 million dollars were invested in duration of 18 years and many academic researchers were supported for this purpose.In 1980s, Algal characterization and strain isolation, studies of algal biochemistry and physiology, process development and genetic engineering made developments/advances for production of liquid fuels from alga; whereas initially, the idea was to produce hydrogen only.

Resource assessments and Techno economic analyses were key aspects of Aquatic SpeciesProgram. Some of the key features of the Aquatic Species Program are described below which have been derived from comprehensive overview of program in 1998;

About 3,000 microalgae strains were collected by the researchers under Aquatic Species Program during a time of 7 years from different sites in northwestern, western and southeastern United States showing a diversity of water types and aquatic environments.

Isolation of strains was carried out frominland saline habitats; which were shallow and undergo swings in salinity and temperature.

Isolated strains were tested for their tolerance in pH, temperature and salinity and ability for lipid production. Then 300 most promising strains of diatoms (Bacillariophyceae) and green algae (Chlorophyceae) were sorted out. In next step, lipid production was studied in the conditions of stress.

This step led to the valuable insights regarding production/biosynthesis of lipids. It was revealed that some species accumulated 20-50 % of dry cell weight as lipids under stress and unfavorable environmental conditions.

Many subcontracts were made under this program in Hawaii, California and New Mexico and an outdoor test facility was operated by the Microbial Products, Inc. in New Mexico. Similarly some other researchers concluded that natural gas which will be derived from algae would not be costcompetitive with gas derived from fossil fuel.

CHAPTER 2

PATHWYS FOR ALGAL BIOFUEL PRODUCTION

CULTIVATION

Assessing the sustainability of production systems for the production of algae biofuels requires examining the different materials and inputs for agricultural production systems to maintain scalability, leverage and minimize costs. Scalable production refers to the ability of a system to maintain algal biomass yields and algae products (mass / hour or mass per unit time) at the commercial scale. System Resistance refers to the ability of a reliable and reliable one to provide a constant output and to avoid failures or malfunctions due to biological crops or physical causes. Costs are the capital and labor costs of a crop system.

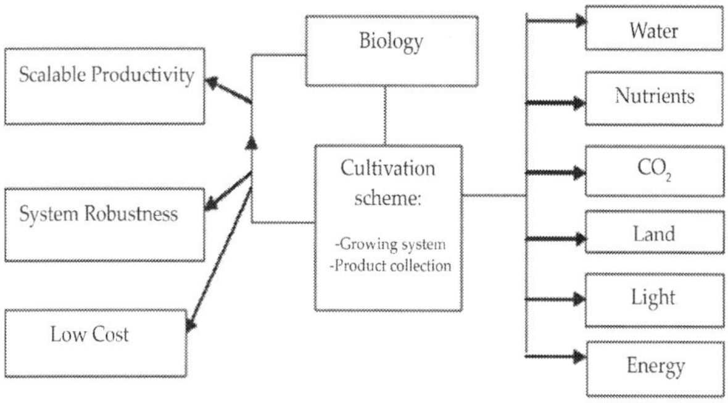

Figure 2.1: Cultivation Schematics

As compared to food crops, algae grow faster and more oil/unit area can be produced. Due to short time for harvesting cycle, algae can be cultivated several times following a different strategy as compared to annual crops. The land which is not suitable for terrestrial crops can be used for the cultivation of algae.

Till date many of the researches for the cultivation of algae focus on the following two types;

Open ponds: this is cheap but chances for contamination are high

Photbioreactors: expensive but clean which may avoid contamination

Open-Pond Systems

Systems for the open pond cultivation of algae have given-up for the algae production with high oil contents. Main flaw of the Aquatic Species Program, as reported by many researchers, is supposed to be the cultivation of algae in open pond which makes the system dependent on the hardiness of the chosen strain. Such strains are required to have characteristics not only to withstand pH and temperature swings but also resilient against bacteria and invasive algae. There are a lot of chances for viral infection in open systems with monoculture. In Marlborough, some trial productions of algae have been taken place in open sewage ponds.

Raceway ponds (Figure 2.2 a-e) they are built as individual units (Figure 2.2b-e) or a unit group without stopping (Figure 2.2a) are tied. Multiple channels allow algae to grow in ponds at a depth of 15 to 40 centimeters. The channels are made of concrete or ground can be coated with compacted plastic. An impeller, propeller or air lift pump always works to vibrate and circulate the mixture to avoid sedimentation of the algae.

Cyanotech: (Figure 2.2c) it has more than 60 ponds, each of which is about 2,900 m^2. The depth of these ponds differs from 30 to 40 centimeters. For a pond capable of obtaining a cellular concentration of dry weight up to 1 gram per liter and productivity were reported between 10 and 25 grams of dry weight per m^2 / day

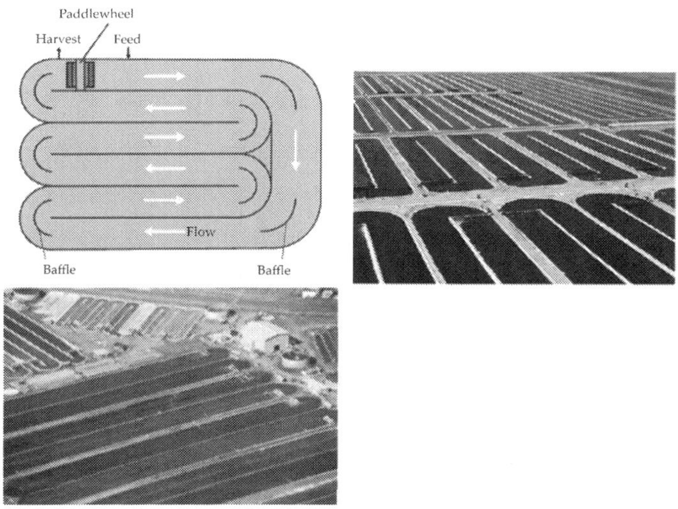

Figure 2.2: (a) Cultivation of Algae in an open pond race way design: (b) Earthrise raceways (c) Cyanotech raceways,

SOURCES: (a) Adapted from Spirulina Source (Spirulinasource, 1999). (b) Earthrise (AlgaeEnergy,2012a). (c)Cyanotech(2012).

Table 2.1 demonstrates the algal efficiency for the open frameworks. There are many variables that may influence the algal efficiency including the open framework sort and the developed algae. In a pool of 100 m2, a cutting edge driven by an electric engine has an utilization of 600 watts (W). It is normal that the aggregate vitality required for sprinters in a lake with an inconsistency coefficient of 0.025 is 20 kW (kWh) per hectare every day for a blending velocity of 15 centimeters for every second and 160 kilowatts of blended day paces of 30 Centimeters for each second.

Different assessments of vitality prerequisites for expansive lakes (e.g 2,900 m2 Cyanotech lakes) extending from around 1200 to 3700 W/ha to blend a speed of 20 to 30 centimeters for every second.

A 85 m2 shaft with a pneumatic course pump has an energy of 195 W in view of compressor proficiency of 70% and air request is 120 liters for each second. Brazil and Chile Panels utilized as an other option to the cutting edge drive engine; It is accounted for that vitality request is just 20 percent of the vitality required for such a challenge cut wheels.

Table 2-1: Microalgae Productivities in Open Ponds

Type of Ponds D	Volume (L)	Microalgal Species	Areal Productivity (gDW/m^2/d)	Volumetric Productivity (g DW/L/d)	Reference
Circular	1,960	*Chlorella* spp *Scenedesmusspp.*	1.61-16.47 2.43-13.52	0.02-0.16 0.03-0.13	Kanazawa et aL (1958) Kanazawa et al. (1958)
Circular		Oscifictoria	15		Sheehan et aL (1998)
Sloped (cascade)	1,970	*Chlorella* spp	25	10	Lee (2001)
Slope	1,990	*Scenesdesmus obliquus*	24.8		Becker (1994)
Raceway		Spirulina *(Arthraspira)*	9-13		Olguin et al. (2003)
Raceway	282	*Spirulinaplatensis*	14.47	0.183	Pushparaj et al. (1997)
Raceway	300	*Anabaena* spp.	9.4-23.5	0.031-0.078	Moreno et al. (2003)
Raceway	135,000	*Spirulina (Arthraspira)* spp.	2-17	0.006-0.07	Jimenez et al. (2003)

Raceway		*Dunaliellasdina*	1.6-3.5		Garcia-Gonzalez et aL (2003)
Raceway	750	*Spirdina plaensis*	15-27	0.05-0.18	Richmond et aL (1990)
Raceway	4,150	*Phaeockroium triconuzwn*	2.4-11.3	0.0028-0.13	Laws et al. (19)
Hybrid system (open ponds and dosed phctobi-creactors)		unknown	30 (anticipated)		Phycal (2011)

SOURCE: Adapted from Chen et al. (2009) Bottom of F

Closed Systems

Lack of structures and equipment required for growth of algae in huge amounts is a big hurdle in large scale production of biofuel. The goal is to maximum use of current hardware and processes of agriculture. Closed systems are the systems which are not exposed to the open air to organisms which may contaminate the culture. The main problem is to find a source of sterile CO2 which may be supplied on cheaper rate. CO2 produced form the smokestack is reported by many researches to be good for large scale production of algae. An idea is to use waste heat and pollutants through cogeneration for production of algae.

Photobioreactors

Concentrated culture of microalga species, without discriminating against environmental benefit, may involve the use of bioreactors. Photoreátricos are transparent pathways or light must be to increase the amount of light available for algal cell packaging crops and have always been mixed to advance nutrient flow and gas exchange. Production of microalgae for photovoltaic has an optimal thickness is about 2-4 cm. Airplanes and flat surfaces are two known types of microalga photoreceptors

Figure 2.3: Tubular Photobioreactors

SOURCE: California Polytechnic State University

All bioreactors have an extensive surface-to-volume proportion (SVR). Because of its wide accessibility, long sun based tubes utilized as essential material. The geometric setups of the tubular photobioreactor involve an extensive variety of even vertical flat triangular propeller lines.

Figure 2.4: External air-lift in Tubular Photobioreactors

SOURCE: Vunjak-Novakovic et al. (2005)

One of the biggest photobiorreactoras foundations on the planet is in a nursery in Klotze, Germany. This establishment comprises of straight even tubes stacked in vertical curves (Figure 2.5). The ability has an aggregate volume of 700 cubic meters (m3), possessing an aggregate territory of 10,000 m2 and produces 35-41 grams of dry weight per m2 every day or 120-140 tons for every year per dry. Keeping divider algae, biofouling, weight drop and pH inclination broke down oxygen or CO2 can happen along the length of the tube. These elements are potential detriments of tubular photoreactor setups can be controlled by an imaginative outline. (Figures 2.3, 2.4)

Figure 2.5: Tubular Photobioreactors, Germany

Flat-plate

Translucent rectangular containers are called Flat photobioreactors which are usually vertical or inclined. These Flat photobioreactors have light path of 1-30 cm (Figure-2.6). Forceful air sparging is used by the Flat photobioreactors to mix the substrate.

Figure 2.6: Flat plate photobioreactors

SOURCE: Algae Energy (2012); NanoVoltaix (2012).

Photobioreactors of algal biomass profitability differ with the kind of geometric arrangement utilized and the algae development (Table 2.2). Numerous inventive generation frameworks have been outlined and are right now a work in progress and testing. New creation frameworks are intended to lessen support costs and comparative watertight upkeep frameworks and keep up high profitability securely and diminish the danger of defilement of shut photobioreactors.

Figure 2.7: Solix Photobioreactor

SOURCE:http://www.agrosustentable.com.ar/Sustentabilidad/sustentabilidadybiocombustibles.html Bottom of Form

These systems contain the Solix, accordion opinion, and the Algenol NASA(NASA)algae growth chart(OMEGA)and the Photon8 aeronautical wave shift system. Photobioreactor is a Solix plastic photoelectric panel (low ratio between height and length). It is designed to cover the vial space to open the pool as a test environment or an inoculated control (Figure 2.7) species. The Solix photobioreactor operation allows the use of water as an open thermoregulation box to open. Aeration and mixing of spray air occurs along the entire length of each panel.

Table 2.2: Microalgae Productivities in Photobioreactors

Photobioreactor	Volume (L)	Microalgal Species	Productivity (gDW/L/d)	Reference
Airlift tubular	200	*Porphyridium cruentum*	1.50	Camacho Rubio et al. (1999)
Airlift tubular	200	*Phaeodactylon tricornutum*	1.20	Acien Fernandez et al. (2001)
Airlift tubular	200	*Phaeodactylon tricornutum*	1.90	Molina et al. (2001)
Inclined tubular	6.0	*Chlorella sorokiniana*	1.47	Ugwu et al. (2002)
Undular row tubular	11	*Arthrospira platensis*	2.70	Carlozzi (2003)
Helical tubular	75	*Phaeodactylon tricornutum*	1.40	Hall et al. (2003)
Parallel tubular	25,000	*Haematococcus pluvidalis*	0.05	Olaizola (2000)
Bubble column	55	*Haematococcus pluvialis*	0.06	Lopez at al. (2006)
Flat plate	440	*Nannochloropsis spp.*	0.27	Cheng-Wu et al. (2001)
Flat plate	100	*Nannochloropsis spp.*	0.30	Rodolfi et al. (2009)
ACCORDION	60	*Monodus subterraneous*	0.40	Cuello and Ley (2011)

NASA OMEGA is a flat panel photoreceptor plate. The front fits the osmous membrane allowing oxygen and water to flow while the photoreceptor is placed horizontally on a water surface (Figure 2.8). OMEGA is designed for the distribution on the surface of salty water bodies (ie sea and sea) to explore cultured blend ripple movements and temperature regulation. This photobioreactor has been redesigned to overcome the reported problems. This design is more complex than previous and significant deviations are included. A vertical series of flat material panels called as Accordion photobioreactor of algae is a continuous or continuous recycling (Figure 2.9) or growth. Adjustable angular vertical planes and adjustable corner planes are designed to increase the incidence of light on the surfaces and improve mixing patterns and flow inside the plates.

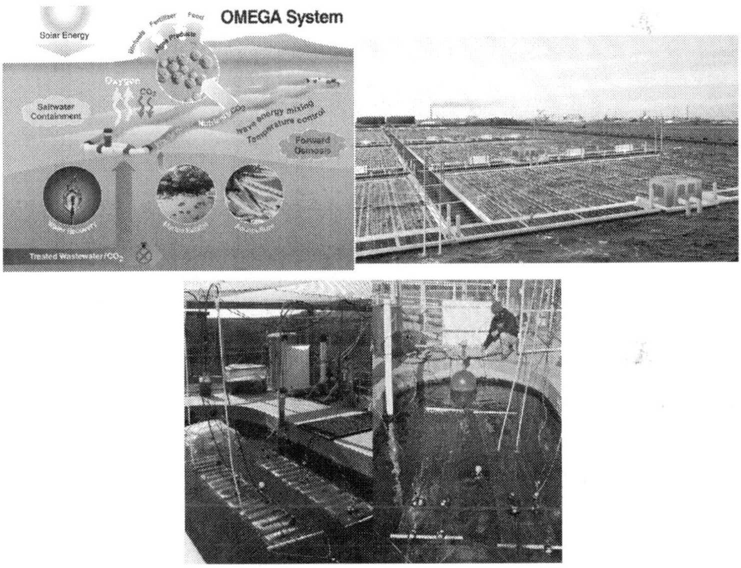

Figure 2.8: NASA's OMEGA photobioreactor

SOURCE:https://www.nasa.gov/centers/ames/research/OMEGA/index.html

A treatment for a sixty liters ACCORDION photobioreactor with flow of liquid of 14 liters/ minute and plate angle of 45° resulted in of 0.30 grams of dry weight/day algal efficiency that was indistinguishable statistically from that of a one liter shake flask control.

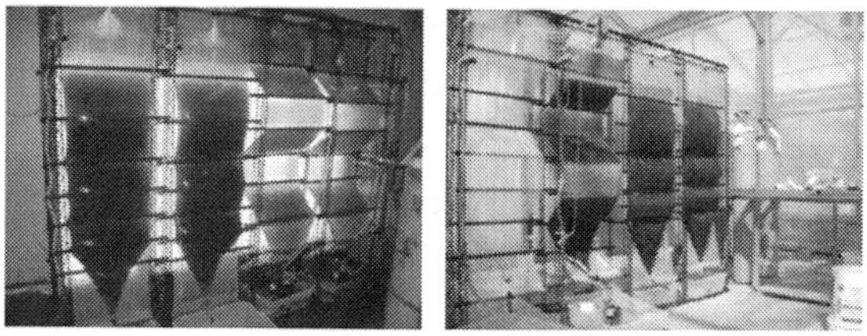

Figure 2.9: ACCORDION photobioreactor.

SOURCE: https://www.slideshare.net/Adrienna/reimagining-the-vertical-farm-2015

The photobioreactor accordion is a modular structure that can be extended by storage modules. This is equivalent to depression pools for scalability, with the addition of drives driven. The PHONOMONIC RECTIFIER is experiencing more than structural design and optimization. The reactor is a horizontal semi-cylinder motor with hydrodynamic plastic Algenol Fotógeno moves back and forth around the longitudinal axis of the photobioreactor for mixing the substrate (Figure 2.10). Photoboreceptor, which is designed for the direct production of ethanol, is used for the cultivation of enolated cyanobacteria enolated by ethanol in an aqueous medium. The ethanol-water mixture is evaporated to form condensation on the photobioreactor substrate and flows to both sides of the inner walls where the plastic sleeves are taken with ethanol-water condensation and transmit a collecting port to one end of the photobioreactor.

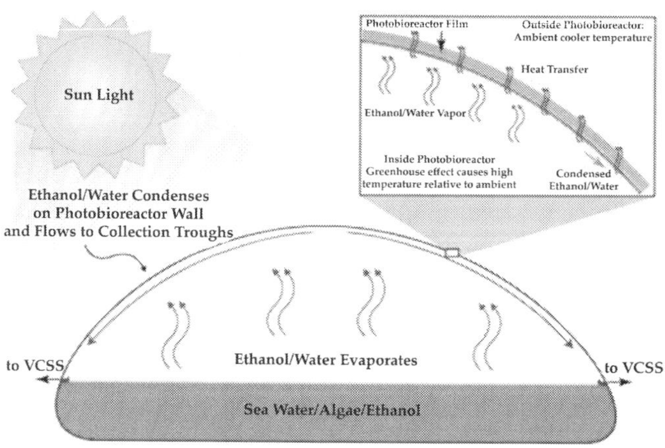

Figure 2.10: Algenol photobioreactor.

SOURCE:http://www.cleantick.com/users/arunesh85/projects/new-and-emerging-technology-in-algae-fuels/18372

Algal Cultivation Systems: Design Considerations

Supplemental CO_2

As the additional carbon dioxide increases significantly the growth rates of algal biomass, integration is predictable as the universal culture of mass culture (Ono and Cuello, 2004a, b, 2006). Yue and Chen (2005) announced, for instance, that the most extreme development rate of the Chlorella ZY-1 strain was gotten when an air stream improved with 10% CO2 (v/v) (Figure 2-11 was utilized And 2-12). The straight development rate under these conditions was around 1.17 g of dry weight per liter and day (Figure 2-12), and the cell fixation achieved 5.77 g of dry weight per liter following 6 days of culture. Improvement rates and cell fixations were higher in societies developed with 5, 20, 30 and 60 percent CO2, with all medications given CO2 2.5 L/min CO2 without combination (Figure 2.11).

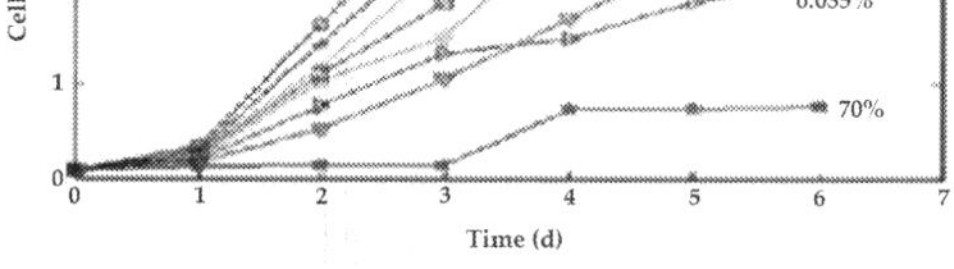

Figure 2.11: Variation in chlorella growth rate (v/v)

Extra carbon hotspots for algal societies incorporate broke up bicarbonate in water or CO2 gas. On the off chance that CO2 gas is provided in open tanks, it is basic to advance the extent of the gas rises to guarantee that CO2 stays in the water and is consumed by algae. Open development with sintered stone or porcelain PVC channels with a line of thin openings in the upper tub. On the surface of the lake, in any case, these strategies include critical CO2 misfortune in the environment in light of the fact that the watery suspension of algae just keeps gas rises for a brief timeframe with the goal that CO2 can't be completely ingested. The achievement of the employment rate was difficult working conditions more than 10% of CO_2 available. In addition, the pipe or duct assembly is technically difficult and porous gas allocation tends to become dirty with algae and other debris and requires periodic cleaning. Researches have shown that the CO_2 production rate of a different open source can be reduced on the basis of pH, alkalinity and ionic strength. It is not a conventional floating CO_2 injector developed in Peru. The device consists of a floating chamber (preferably sealed tubes and covered with transparent PVC sheets) of hollow quarry. Flotation gives the gas pillow below the lid. Through a ball valve framework, the compartment is loaded with unadulterated CO2. At the point when the gadget rises through its lightness, the valve is close and the other way around. The resulting loss of CO2 begins at 4 percent of the aggregate CO2 supply sometimes.

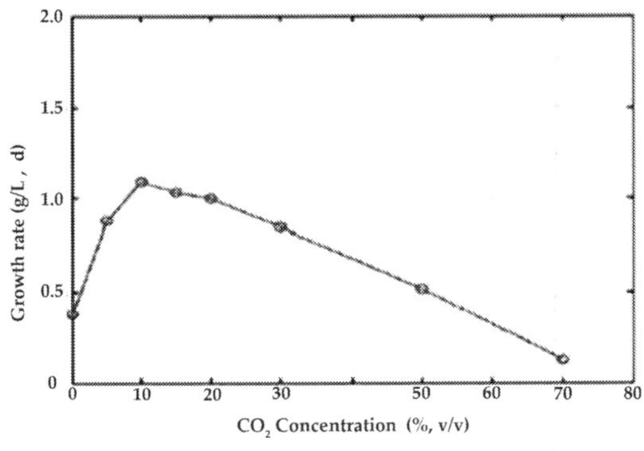

Figure 2.12: Variation in chlorella growth rate (v/v)

The amount of CO2 release of bioreactor gas depends on the specific type of photobioreactor and how it works. Plain plate photovoltaics, for example, generally have much shorter distances than tubular photobioreactor gas. Unlike the open track tread, it can minimize the CO2 emissions of effluent gas recycling cycles. If the flow of recycled gas has a high concentration of oxygen, it is necessary to address the challenge of eliminating oxygen inhibition to control growth of crops.

Contamination and Stability of the Algal Culture

The development and support of monocultures is essentially basic, if the yield of the item (e.g., algae oil or ethanol) relies upon the development of an especially refined algae species. Refering to Moheimani and Borowitzka (2006), an exploration investigated that some open tank frameworks could keep running for a half year without noteworthy levels of sullying. Be that as it may, keeping up strength and high profitability of biomass developed species by a solitary gathering of nonaseptically open algae stays a standout amongst the most impressive supply outside parks. Monocots, huge algae monocytes have a tendency to be attacked by parasites and pathogens and the insurance of undesirable harvests is a considerable supportability of algae. The primary sort of contamination is algae culture including algae, microscopic organisms, algae, infections, zooplankton, parasites and bugs. These sorts and areas of contaminants in algal societies rely upon the encompassing ecological conditions, particularly algal societies and particular culture frameworks being used.

Contamination of algae is inescapable, in light of the fact that the development conditions are reasonable for the types of unavoidable gets of algae and other nearby species. Indeed, even on account of Spirulina cyanobacteria developed under high alkalinity and profoundly segregating high alkalinity pH, defilement by other cyanobacteria or swaying algae has been accounted for subotimales conditions less than 15 grams of bicarbonate in liters development. In outrageous cases, intrusive types of algae could wind up noticeably predominant and attack wanted species. In the way of life of Scenedesmus spp., regular contaminations incorporate Chlorella spp., Selenastrum spp., and a few types of diatomas. In one of the investigations did in Japan, it has been accounted for that keeping up the Chlorella inoculum collect is important to direct the begin of the treated societies. Albeit other algae cannot be expelled from an open tanks framework, methodologies can be altered to keep up tainting at satisfactory levels and the utilization of high convergences of wordy immunization culture framework cleaning.

Monas polluted with algae spp. Or, then again other protozoal species are frequently totally pulverized inside 12 to 18 hours after the debasement is first recognized. Chytrid half breeds have been distinguished in a few algae societies and frequently exhibit plagues that occasionally result in total harvest misfortune. The Chytridium spp. it is the most risky for Chlorophyceae parasite culture, and frequently shows up with the zooflagellate Aphelidium spp. Scenedesmus Infections the way of life of these creatures have been observed to be basically all inclusive. This contamination is portrayed by an extreme flocculation of algae delay, a culture medium and dark colored shading decreased oxygen improvement. At the point when these side effects show up in the algae culture, it is past the point where it is possible to control the parasite. The main Aphelidium natural control disease at a beginning period is a weakening of ordinarily developed with new soil, enabling the algae populace to stay in an exponential development stage so their rate of augmentation surpasses the frothing operator. Scenedesmus spp. what's more, Chytridium spp. they were treated with organic product in the use of Benomyl Fungicide Israel (carbamate methyl-1-butylcarbomoylbenzimidazole) at a measurements of 1 mg for each liter (mg/L). C. reinhardtii parasitic pollution can be controlled utilizing carbendazim fungicides (1 mg/l), methyl thiophanate (20 mg/l) and benomyl (20 mg/l). It has additionally been demonstrated that a mix of carbendazim and ampicillin cefotaxime and dispense with or decrease defilement of C. reinhardtii by three distinct microorganisms and growths tried two unique.

There were periodic transmission reports of zooplankton Lycrymanis spp., Colpidium spp. what's more, Vorticella spp., in spite of the fact that these offices have effectsly affected algae development. Sullying by a gathering called rotifères Branchionus, be that as it may, can keep the development of algae cells and in more serious cases it might harm the entire ocean growth collect. The most helpful control is not as much as the pH culture at around 3.0 by expansion of corrosive and nonstop culture up to pH from 1 to 2 hours before the pH is shifted back to 7.5 with potassium hydroxide (KOH). Rotifères treatment viably evacuates no unsafe algal societies.

Operations in Open-Pond

Open-lake administrations for the vast scale fabricate of algal biofuels likely should be overseen as troublesome, bioengineered frameworks by applying known standards of populace, group, and biological system environment. Different setups for open-lake frameworks can be imagined, two of which are talked about underneath.

Hybrid system for Single species (target strain inoculums)

A hybrid outline limit with respect to algae development includes the utilization of a photobioreactor front later connected to open lakes. The photobioreactor framework would be a reproducer giving a high algal target inflow for creation frameworks and directing downstream channels. Be that as it may, a few key environmental components can conceivably obstruct long haul stable half and half single voltage or single voltage in all frameworks utilize open tanks.

Right off the bat, probably single voltage mixture frameworks would ceaselessly be subjected to potential microfauna and microflora attacks show in the neighborhood scene. For instance, sit out of gear stages or ordered people can be stored on the surface of open lakes by different gatherings of cyanobacteria and algae by coordinate statement of barometrical particulates or in relationship with rain and snow. Away, open tanks will be attacked by the group of different water buyers, including rotiferi, ciliati, bug hatchlings, shellfish and zooplankton. These creature trespassers can be transmitted for the most part from creepy crawlies, transitory sea-going flying creatures, and other territorial rotatable creatures and by wind and rain. Ordered personalities and the aggregate number of species in the long run move toward becoming inhabitants of these lakes relies upon the size and structure of local species allelopathic relationship with wanted species, asset rivalry and profitability and surface lake frameworks. Some herbivorous intruders are exceptionally undesirable in light of the fact that their boundless development can enormously lessen algae development.

Second, the principle reason for the upstream photobioreactor is to give a steady supply of the objective algal strain. Be that as it may, the objective algal strain may not keep on dominating the algal group, once it is attack by cyanobacteria and algae in the downstream open lake. The presence of such issue relies upon whether the objective strain is hereditarily altered or normally happening. Such intrusions could bring about the abatement of the general biomass yield, thus requires expensive interferences of biomass creation for completing framework, cleaning, and strain re-foundation.

In the most situations where the cross breed development framework is in operation, the objective strain would be profoundly forceful and would continually remain the overwhelming essential producer in the lake. Be that as it may, solid fleeting elements in species structure are seen in many polycultures. Singular species tend to show real changes in relative incredible amount as conditions change over natural time. Once in a while various animal groups increment from about untraceable plenitude to solid numerical predominance amid a time of just weeks. Such solid species elements are undesirable in light of the fact that the swaying of target strain wealth could make solid flimsiness in algal biofuel fabricate rates by means of changes in algal biomass, lipid content and atomic arrangement, lipid extractability, and lipid collect capacity. Richmond and Zmora (2003) abridged the generation of Nannochloropsis for rotifers and as an immediate aquaculture encourage.

The most noteworthy lipid content in Nannochloropsis was accounted for by the scientists in summer; rather than this eicosapentaenoic corrosive level was most elevated in winter. Gathering of culture in 25% - 30% at everyday schedule was the most noteworthy biomass revealed amid summer. For this situation low chlorine levels and pH shifts were utilized to control the contaminants.

Mixed species systems

Another outline of algae plant using vast water frameworks is vaccinated with planktonic agglomerates acquired from regular water bodies in the close-by landscape. Such vaccination division contains a delegate gathering of local types of smaller scale fauna and microflora. This gathering of sea-going species contains infections, microscopic organisms and parasites; Cyanobacteria and microarray eucariotics and macroalgae; And water customers, for example, rotiferi, ciliati, creepy crawly hatchlings and zooplancton shellfish. An arrangement of blended animal categories gives potential advantages of better biomass creation and more prominent culture security. Rural writing has for quite a while demonstrated that the accumulation of a few types of products (polyculture) ordinarily gives higher rates of a solitary harvest biomass. This marvel is known as “finished execution”. There has been extreme biomass effectiveness on algae gatherings, prompting Smith et al. (2010b) for the speculation that blended harvests species could create higher yields of biomass and lipid algal agribusiness of a solitary animal varieties. A specialist tried this speculation both regular microalgae groups as in the lab and found an expansion in lipid generation in different algae groups as respects single-developing algae become under a similar asset supply. This examination bolsters the proposal that microalga groups that exist in nature and that develop in open tanks could possibly store more sun powered vitality than groups of a solitary animal categories developed in shut photobioreactors. The fuse of the environmental advantages of the utilization of asset assorted variety in algal biomass generation could give an advantageous approach to build the execution and quality of developing algae for the creation of biofuels.

Another key biological cooperation that applies to the creation of algae on business scale in open tanks is savage prey flow. Like their common analogs, simulated lakes open frameworks build up a few natural groups. Particularly in light of the fact that they contain essential makers (algae) and essential customers (herbivore zooplankton), these lake groups tend to indicate savage swellings of prey populaces like flare-ups of bugs that create harm to crops in earthly agribusiness. Specifically, the uncontrolled development of huge herbivorous zooplanktons, for example, Daphnia magna or Daphnia pulex, is undifferentiated from embeddings abundance nibbling in a field. They can defeat algal cells and prompt decreases in the request of extent of algal biomass yields. Particular brushing of attractive species alongside the arrival of supplements from herbivores can adjust the creation of occupants expanding levels of undesirable algae levels. From the perspective of harvest security, this is an exceptionally undesirable result. Notwithstanding, oceanic nets can be modified to diminish algal biomass wounds through a best control. For instance, predatory fish (Gambusia affinis) might be extra to expansive lake creation frameworks to expel zooplankton and boost algal biomass developed in lakes for the generation of biofuels. The blended group can have a nutritious esteem that shifts with changes in ordered creation. Desires of variety in structure adjust the lipid content and the potential nature of algal biomass for them.

ALGAL BIOMASS PROCESSING INTO FUELS

Another outline of algae plant using untamed water frameworks is immunized with planktonic agglomerates got from common water bodies in the adjacent landscape. Such immunization division contains a delegate gathering of local types of smaller scale fauna and microflora. This gathering of oceanic species contains infections, microscopic organisms and parasites; Cyanobacteria and microarray eucariotics and macroalgae; And water purchasers, for example, rotiferi, ciliati, creepy crawly hatchlings and zooplancton scavangers. An arrangement of blended animal categories gives potential advantages of better biomass creation and more noteworthy culture steadiness. Farming writing has for quite a while demonstrated that the gathering of a few types of yields (polyculture) ordinarily gives higher velocities of a solitary harvest biomass. This wonder is known as "finished execution". There has been over the top biomass effectiveness on algae gatherings, prompting Smith et al. (2010b) for the theory that blended harvests species could deliver higher yields of biomass and lipid algal horticulture of a solitary animal groups. A scientist tried this speculation both characteristic microalgae groups as in the lab and found an expansion in lipid creation in different algae groups as respects single-developing algae become under a similar asset supply. This investigation bolsters the recommendation that microalga groups that exist in nature and that develop in open tanks could conceivably store more sunlight based vitality than groups of a solitary animal categories developed in shut photobioreactors. The fuse of the biological advantages of the utilization of asset decent variety in algal biomass creation could give a helpful approach to build the execution and quality of developing algae for the generation of biofuels.

Another key biological communication that applies to the generation of algae on business scale in open tanks is ruthless prey elements. Like their common analogs, fake lakes open frameworks build up a few organic groups. Particularly in light of the fact that they contain essential makers (algae) and essential purchasers (herbivore zooplankton), these lake groups tend to indicate ruthless swellings of prey populaces like episodes of creepy crawlies that create harm to crops in earthly farming. Specifically, the uncontrolled development of substantial herbivorous zooplanktons, for example, Daphnia magna or Daphnia pulex, is similar to embeddings overabundance munching in a field. They can defeat algal cells and prompt diminishments in the request of extent of algal biomass yields. Particular brushing of alluring species alongside the arrival of supplements from herbivores can modify the piece of occupants expanding levels of undesirable algae levels. From the perspective of harvest wellbeing, this is an exceedingly undesirable result. Nonetheless, amphibian nets can be changed to decrease algal biomass wounds through a best control. For instance, savage fish (Gambusia affinis) might be extra to vast lake creation frameworks to expel zooplankton and boost algal biomass developed in lakes for the generation of biofuels. The blended group can have a healthful esteem that shifts with changes in ordered piece. Desires of variety in sythesis modify the lipid content and the potential nature of algal biomass for the fabricate of fuel.

Culturing of microalgae aqueous medium

What's more, lipids in the algae are in like manner non-miscible water. On a fundamental level, the detachment of algae oils from watery culture media can be unconstrained and requires little vitality. By and by, aalgae is isolated from the development medium and lipids are isolated by algal biomass properly at high vitality utilization. Lessening the vitality required can be accomplished through enhanced algae strains of specialized changes, and by a positive connection between the two. For instance, algae improvements that expansion cell culture thickness on a basic level lessen the measure of water to be expelled amid recovery. Lessening process water, the various things would decrease the vitality expended amid algae collecting.

Reaping techniques change impressively relying upon whether or macroalgae microalgae are refined. In spite of the fact that the concentration of this proportion is microalgae, macroalgae are mechanically collected with generally low vitality contributions of today. Macroalga gathering is an area of more than 400 years, with the advancements that are as yet being proposed to enhance productivity? For microalgae the techniques is utilized for collecting by dimensional avoidance or thickness based detachment. These incorporate filtration, centrifugation, sedimentation, buoyancy, flocculation (coagulation) and electrophoresis strategies. Flocculation and sedimentation of gravity are comparative. Common thickness division can be quickened by including operators that reason totals of microalgae. Total triggers incorporate pH varieties or other substance triggers. The algae can settle or buoy to the surface of the fluid. Partition is a generally low vitality decantation of a watery stage and aalgae control stage. (Engineered) inorganic or natural flocculants are utilized, with the nature and aura of these material manageability concerns.

The utilization of inorganic flocculants (for instance, alum and ferric chloride) cause ecological issues and change. Flocculants go through the algal biomass process and should be represented all the while. Hoisted leftover metal fixations in algal biomass confine use as co-items for wellbeing reasons. Natural flocculants might be powerless to anaerobic processing by expelling them from the reuse stream. The nearness of flocculants may impact fitting uses for algal biomass as co-items.

Centrifugation and filtration methods can be utilized alone or in mix with a preparatory thickness partition. Centrifugation technique concentrates offices rapidly however requires a high cost of capital and working expenses. Filtration can be inadequate in light of the fact that the minute algae cells tend to stop up the channel. Centrifugation and filtration are getting significant consideration, and developments are accounted for. They have audited different strategies for gathering, for example, acoustic control of algal cells or electrophoresis systems, including electrolytic and electrolytic flocculation coagulation, yet the accumulation rate and unwavering quality shift. As noted before in this section, they likewise tried endeavors to grow hereditarily designed algae consolidating attractive nanoparticles to diminish vitality costs for accumulation and parchedness.

Table 2.3: Characteristics of Microalgae Harvesting Techniques.

Harvest Methods	Suspended Solids Concentration (%)	Operating Costs per Gallon of Water	Cell Harvesting Efficiency	Algal Species
Centrifuging	High (< 22%)	Very high ($20 to $50)	>90%	Almost all except the very fragile
iltration/screening	Medium to high (5% to 18%)	Medium to high ($10 to $20)	20% to 90%	Algae with large (>5 μm) cells
Flocculation	Low to medium (3% to 6%)	Low to medium ($3 to $10)	50% to 90%	Algae with low density
Bioflocculation	Low to medium (2% to 5%)	Low ($0.20 to $0.50)	About 90%	
dimentation/settling	Low (0.5% to 3%)	Low to medium ($0.50 to $1.50)	10% to 90%	Algae with high density

Extraction Technique

In the event that lipids are considered as the essential wanted item, the oil ought to be extricated upon the harvestation of the algal biomass. For handling strategies utilizing entire cells, culture could be all that is required for the following phase of fuel generation. Biodiesel generation is an innovation that in many variations require the gathering of algae lipids for additionally preparing. Extraction of algal biomass oil is ended up being troublesome. Dissimilar to oilseed oil recuperation, there is no all-around characterized economically accessible lipid algae extraction process in the market. Similarly as with seed oil recuperation, extraction with natural solvents has been tried with algal frameworks. There are two contrasts between the extraction of oil from seeds and algae. To start with, straightforward crushing of sleek seeds makes a removable nourishment. Second, oleaginous flour is around 90 percent strong. Interestingly, algae biomass is high in water, and effortlessly broken cell films to deliver effectively removable oil.

They have revealed lifted levels of heterotrophically gathered algae oil articulation. It is said that the levels alluded to up to 80 percent are high to the point that lone cell lysis is required for gathering, staying away from extraction. There are no reports that photoautotrophic culture has achieved such high portions of oil.

Oil extraction should be possible with dried kelp or wet cake accumulation. Drying is high-power vitality however delivers a material that can be mechanically treated to open access to oil extraction. Once dried, the oils are separated. Attractive dissolvable elements incorporate high dissolvable power, low lethality, low particular warmth, low vaporization warm, minimal effort, high accessibility, and ideally non-combustible. Cases of solvents incorporate hexane (most generally alluded to as the dissolvable utilized for algae extraction), methanol, ethanol, butanol, ethyl acetic acid derivation, chloroform, oil and ether. The dissolvable is dissipated; algae oil is recuperated and reused. Solvents recuperation is high yet not 100 percent. It can prompt lost 0.5 to 5 percent. The extensive volume of dissolvable and dissolvable vapors in the process can be a danger of flame and blast.

Solvents utilized for extraction are normally wet immiscible water sparing in vitality utilization in dissolvable recuperation. Biomass is by and large handled as entire cells on the grounds that mechanical layer breakage is troublesome in a wet batter. Dissolvable lipid arrangement algae are recuperated for sanitization. The dissolvable and algae oil are dissipated, as in the treatment of oilseeds.

Combined Harvest and Collection

Different techniques have been produced attempting to utilize electric fields, warming or different intends to scatter ocean growth oil without collecting the algae. Oil Origin, Open Algae, and Diversified Technologies are among the organizations that try to considerably decrease the vitality gather and oil accumulation needs. The start of these strategies is fueled by developing algae to break cells (lys). Lipids immediately isolate biomass cells while expanding the biomass surface sinks. The normal outcome is a strong pellet, a watery layer and a layer of free oil with the goal that partition for straightforward, productive and effective vitality gravity will recover oil. These frameworks offer the guarantee of a considerable decrease in vitality utilization and taking out the utilization of solvents. Another reap of extraction method, helped microwave drying, misuses the microwaves to energize the fluid inside the algal cells, causing the cell divider to break and discharge lipids. The solvents are then used to separate the lipids. The helped microwave drying stove can spare time on the grounds that algal biomass does not need to be dried out before lipid extraction. It is not known whether microwave warming reductions the lipid quality or scale comfort to the amounts required for fluid fuel transport. Presently, there is little data on the viability and vitality adjust of these procedures.

In-Situ Esterification of algal biomass

Treatment of wet algal biomass with alcohols can bring about the immediate gathering of unsaturated fat methyl esters. In any case, this innovation has so far demonstrated a little advantage when contrasted and the more customary oil recuperation advancements.

Processing

The vast majority of the introductions to this council have distributed writing on the creation of powers got from the generation of lipid microalgae. Nonetheless, they don't speak to every one of the choices accessible for the treatment of algae in fuel. Many handling alternatives are talked about however not depicted in the writing, making it hard to look at. There are three fundamental parameters that influence the biofuel energizing algae (Figure 2.13):

1. Micro versus large scale i.e., development os miniaturized scale algae or full scale algae for the fuel creation
2. Short length versus stable conditions i.e., development of the algae for brief term and gathered for preparing or development under stable conditions where wanted items radiated by the algae are consistently reaped
3. Biological versus substance forms i.e., the utilization of biochemical or thermochemical procedures to deliver biofuels from algae.

In spite of the fact that the supportability challenges recognized in this report can't be tended to every now and again without a quality information, in any case, the ordering the algal biofuel generation pathways by these three key parameters enables some broad remarks to be made about every blend of parameters (Figure 2.13)

The classes used to recognize treatment rely upon the idea of algae if a steady culture is utilized and the post-treatment is concoction or natural. Algae can be subdivided into microalgae and full scale algae, if algae are types of crisp or salted water. The progression of culture is essential and to a great extent isolated into whether the want is for a steady culture after some time or if algae are killed in collecting the item. This assignment is consequently to a great extent dictated by the way that the coveted items are kept up inside the extracellular or algal cell dividers. At present all procedures portrayed utilizing stable societies and radiating attractive items utilize shut photobioreactors. Introductions to the panel raised questions with reference to whether extracellular items could be gathered in open framework tanks because of potential microbial utilization of the item. Ordinarily utilized substance forms, including ordinary oil extraction and small scale algae handling oil at biodiesel. They have exhibited organic procedures, for example, microalgae biomass maturation.

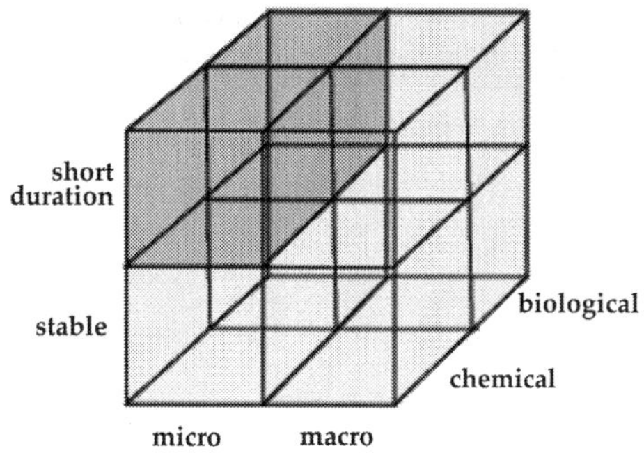

Figure 2.13: Matrix representing combinations of important parameters that explains pathways for algal biofuel processing

NOTE: The gray boxes indicate combinations of pathways that are not pursued, to best of the committee's knowledge.

Microalgae Harvesting for Chemical Processing

The degummed oil is handled in different ways. Two principle items are usually said: biodiesel transesterification and conventional diesel or green diesel call. In the generation of conventional biodiesel, methanol, and a synergist base, it responds with triacylglycerol algae to deliver a methyl greasy ester. Normally homogeneous impetuses, sodium or potassium, are supplanted by heterogeneous impetuses, diminishing waste. Glycerol is delivered as co-created in the two strategies. Late patterns have been towards the creation of hydrogenated diesel as opposed to outside. In pressure driven preparing, hydrogen responds with raw petroleum to deliver algae, propane, and water. Diesel is dealt with more like diesel and is said to offer preferable execution over outsiders. It is assumed that hydrogenated diesel motors are perfect with existing oil framework and in the event that they are fundamentally the same as oil fills are considered outflows.

Extracellular Secretion of Products by Microalgae for Chemical Processing

Algenol and Joule advances are two organizations that adventure the capacity of algae to discharge extracellular items. The items are collected in culture medium for resulting recuperation. Algenol confronted the exam and depicted his strategy for delivering magazine articles and licenses. Algenol specifically utilized cyanobacteria to create ethanol. Joule Technologies has licensed cyanobacteria that create specifically alcano. Eukaryotic living beings have additionally been portrayed; Algae and cyanobacteria radiate a scope of materials that could be utilized for fuel creation. Better depicted in the distributed writing process is the Algenol technique to create ethanol. Ethanol just requires sanitization and does not require additionally handling.

The creation strategies portrayed utilize shut photobioreactors. This is likely an aftereffect of the want to keep up stable harvests for drawn out stretches of time. The presentation of species in rivalry with algae or microbial tainting would be unsafe. Shut frameworks are a way to guarantee the immaculateness of the way of life and the consistent nature of the item

On a fundamental level, the vitality required for division in this method of operation can be low. Designed living beings communicating miscible items would prompt the unconstrained partition that does not require sustenance. Stable yields can be kept up that require least water outlets. Water is plainly expected to renew any loss of photosynthesis and water amid handling. While the guarantee is awesome, the distributed examinations are lacking for exact evaluation of aggregate vitality and LCAs to be performed.

Microalgae Harvested with Product Collected for Biological Processing

Aging of biomass microalgae is presently a work in progress on a business scale. Microalgae give carbon sources, for example, proteins and sugars that can be abused by aging. A few advantages incorporate quick development rates, short collecting cycles, and nonattendance of lignin. Like different biomass maturations could deliver an extensive varicty of items. There is an absence of inside and out examination on this way elaboration.

Extracellular Emission of Products by Microalgae for Biological Processing

Proterro depicts a strategy that utilizations microalgae on a photobioreactor to create sugars that can be recouped additionally utilized as a crude material for different maturations. Sugar can be utilized for any maturation procedure for the generation of energizes. Ethanol is presently the biggest volume of fuel delivered by maturation. Butanol, farnesene, alcohols, and different items can likewise be created by the aging of sugars. Coordinate generation of sugar requires algae development is ensured against artful microorganisms and requires an ecological seal photobioreactor.

Macroalgae Harvested with Product Collected for Biological Processing

Bio Architecture Labs has as of late declared the advancement of an innovation for maturation of macroalgae biomass. The utilization of macroalgae as a crude material for biochemical transformation is conceivable by the improvement of living beings fit for using alginic polysaccharides. Transport elements and alginic digestion have been intended for the combination of ethanol. This permits the immediate combination of ethanol from macroalgae. It is said that darker macroalga are fascinating wellsprings of energy that don't require the contribution of manures, farmland and freshwater assets, and in this manner are not in rivalry with the current harvests for those assets. Sugars can be discharged in basic mechanical operations, for example, crushing and granulating, in light of the fact that macroalgae don't contain lignin. Moreover, development strategies are built up on the grounds that macroalgae are gathered for nourishment fixings, encourage and composts. Saccharine japonica turned out to be an aging substrate. Gracilaria Salicornia has additionally demonstrated an appropriate substrate for microbial maturation of ethanol.

Microalgae or Macroalgae Harvested for Whole-Biomass Processing

There is an expanding enthusiasm for biomass transformation for the whole biomass preparing industry. Pyrolysis of the whole biomass delivered an updateable biocrome. A current audit demonstrates that these courses have money saving advantages contrasted with different biomass transformation innovations. The created material can possibly be utilized as a part of a current refinery, sparing capital over different alternatives.

We have investigated different types of pyrolysis. It portrays the hydropyrolysis is reasonable for use in the treatment of algal biomass. It is said that the low substance of sweet-smelling mixes in algal biomass (because of the nonattendance of lignin) causes algal biomass is an especially decent beginning material for hydropyrolysis.

It has revealed the advancement of two option pilot scale, both the generation of non-renewable energy source perfect materials. These engineered fats are pronounced good with existing refineries. In one process manure is a co-item. There is little documentation for a nitty gritty mass and vitality adjust of the procedure. It can be expected that itemized biomass studies will create comparative outcomes when algae are the crude material. That is, the preparing of the whole cell gives a conceivably suitable fuel substitution condition, misusing the current refinery framework to lessen dangers and expenses.

Status of Algal Biofuel Production:

- The creation of algae biofuels is quickly advancing, and thusly, any state relationship is out of date at finishing. Presently, there are no biofuel generation offices equivalent to the US normal biodiesel refining limit, around 13.5 million gallons (51 million liters) at yearly conditions, significantly less matching with the bigger than 378 million liters for each year (NBB, 2012). Many activities are still under innovative work and it is hard to get correct creation numbers.
- The nearness of photosynthetic and synthetic definitely extraordinary efficiencies among algae species features a basic requirement for essential and connected research to extend germplasm range accessible to the organization
- Expanding the light range which is valuable for photosynthesis.
- Improving the dispersion of episode light to different amphibian photosynthetic scale-up forms.
- Enhancing the proficiency of Rubisco or other fundamental physiological procedures to better use carbon could prompt sensational changes in efficiency.

Extra new leaps forward in ranges, for example, the ability of algae to change over supplements into biomass all the more proficiently or the decrease of handling costs related with collecting and dewatering (for instance, by means of hereditary upgrades that support auto flocculation) additionally can possibly additionally enhance the vitality adjust and to improve the general maintainability of an extending algal biofuel industry.

Table 2.4: Major Inputs and Outputs for various Algal Processing Technologies.

Duration of Cultivation System	Type of Algae	Type of Processing	Product	Major Inputs for Processing	Coproducts
Short	Microalgae	Chemical	Hydrogenated diesel	Hydrogen, extraction solvent	Algal biomass, propane
Short	Microalgae	Chemical	Biodiesel	Methanol, base catalyst	Algal biomass, glycerol
Stabe	Microalgae	Chemical	Ethanol		
Stabe	Microalgae	Chemical	Alkanes		
Shrt	Microalgae	Biochemical	Ethanol, other products		
Shrt	Microalgae	Biochemical	Ethanol, other products		
Shrt	Microalgae	Chemical	Pyrolysis oil		Char, off-gases
Shrt	Microalgae	Chemical	Pyrolysis oil		Char, off-gases

Similarly imperative is examine on edit insurance, which concentrates on decreasing biomass misfortunes for pathogenic and herbivorous specialists, for example, defilement by other algae species, is generally inescapable, particularly in open algae crops, enhancing comprehension of frameworks of existing algal biomass creation. Bioengineering frameworks would be helpful. This can be accomplished to some extent by applying the standards of populace, group and biological community (nature). It would likewise be helpful to recognize qualities and eco-physiological parameters (bioengineering) to give better insurance against herbivorous and pathogenic levels in business scale generation. Physical procedures and upgrades in the algae development techniques utilized for reaping, drying and changing over algal biomass powers are so vital for the supportable advancement of algae biofuels as enhanced algae strains. They can investigate new strategies to lessen vitality prerequisites for the transformation of algae developed into a dried out watery arrangement that would then be able to be transformed into fuel. Innovative work in the comprehension of dry algae can be changed into a fuel and if algae can create a valuable hydrocarbon straightforwardly without the need to gather and get dried out and insignificantly treat it could be a critical commitment to decreasing generation costs. Fundamentally, the inquiries concern the combination of science, biology and comprehension of designing and efficient change of far reaching biofuel algae.

PATHWAY FOR ALGAL BIOFUEL PRODUCTION PROCESS

The entire procedure of collecting from algae yields and gathering, lastly the change of fuel is known as the street to the creation of algae biofuels. The aim is to bunch pathways that offer normal procedures to help neutralize asset prerequisites and the effect of various methodologies and to help clear up key natural building abilities and need to enhance manageability advance.

Characteristics of Biofuel Pathways

The pathways for fluid biofuels share numerous normal qualities regardless of the biomass crude material utilized. Everybody has a development stage, a gathering or accumulation stage and completing or step preparing. Some land crops are utilized or considered for use as biofuels in view of its capacity to deliver oils, their capacity to create starches is effectively changed over into fuel by microbial activity or carbon catch ability with low assimilation. Oleaginous deliver, for example, soy, jatropha and camelline, gathered and the oil is isolated for additionally handling. Sugar genuine sweetener and maize starch can be effectively changed over to yeast and bacterial ethanol. Vitality harvests, for example, poplars, grass change and Miscanthus are picked as a result of their low development supplements or their capacity to store carbon in the dirt. Lignocellulosic biomass can be changed over synthetically or naturally to thermo-substance fluid powers. There are frameworks like algae biofuels for each of these sorts of crude materials. Investigation of choices in algae development, collecting and handling is confused by the expansive number and many-sided quality of choices. Nonetheless, ways can be gathered by the principle attributes they share that are influencing the utilization of assets and vitality adjust. Some illustrative pathways are examined to show the present condition of the innovation and where advance is expected to diminish asset necessities.

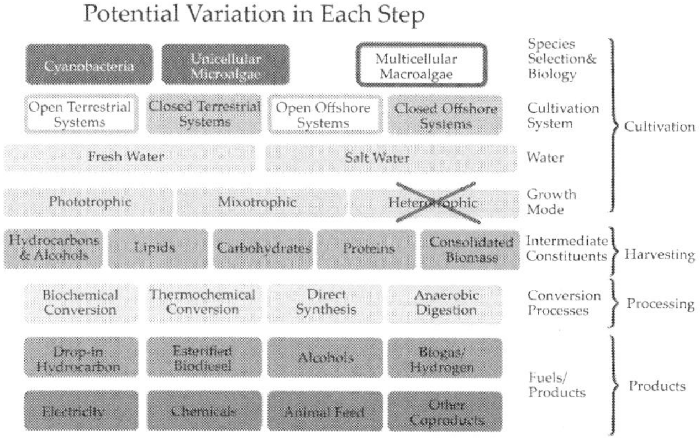

Figure 2.14: Ways to cultivate and process algae in to fuels and their products.

NOTE: Heterotrophic routes are outside the scope of this analysis.

It is likely that patterns in science and innovation for the creation of different biofuels delivered in the generation of algae biofuels, as it creates as industry. These patterns incorporate enhanced biomass creation and handling, and enhanced relative lifecycle affect examination and necessities for different wellsprings of option fluid energizes through lifecycle appraisal. A further pattern is the move towards fuel misfortune that is perfect with the current oil fuel framework. Ethanol and methyl esters of unsaturated fats (ordinarily called FAME or biodiesel) have similarity and execution issues in vehicles that hamper their appropriation. Current patterns are moving towards the generation of unadulterated hydrocarbons or fuel-perfect fuel foundations and existing blend vehicle innovations.

Fuel and vitality generation from algae is not a universal area and has proposed an assortment of creation frameworks. Figure 2.14 is an improved plan intended to confine possibilities and gathering the different biofuel periods of algae generation chain. Each line in the graph subtle elements a procedure step or a handling alternative. A few blends of development and handling choices have prompted more than 60 distinct strategies proposed for the creation of algae biofuels.

Culture is heterotrophic, as indicated by outline, outside the extent of this report. The prohibition of heterotrophic creation of this relationship is not a judgment on the legitimacy of these methodologies, yet an impression of the need to consider photosynthetic algae as a crude material for the generation of fuel. This report takes a gander at approaches to deliver powers for the vehicle of ocean growth fluids. Era of gas and hydrogen creation is not talked about. Proteins are just considered as co-items. Streets help delineate asset prerequisites and potential effects related with the size of various ways to deal with creating biofuels from algae. Diverse ways additionally enable you to look at the changed directional methodologies, enabling you to reach determinations, recognize entanglements and create conceivable arrangements. The reference track is taken from the current specialized and financial investigation of the national creation of the Renewable Energy Laboratory of algae biofuels.

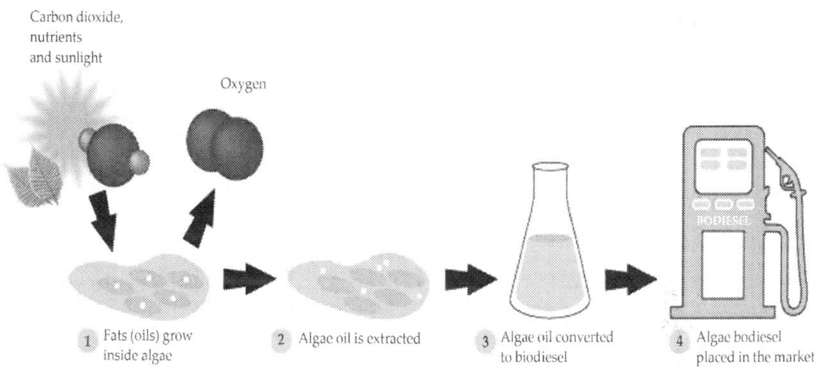

Figure 2.15: A simple representation of steps involved in processing algae to Bio-fuels

SOURCE:http://www.bloggang.com/viewdiary.php?id=surya21&month=042014&date=24&group=3&gblog=204

Reference Pathway-Raceway Pond Producing Drop in Hydrocarbon

The reference path meant that microalgae are grown in salted water in an open pond. The algae are harvested and lysed to release the lipids that are harvested for further green diesel processing (also called renewable diesel), a hydrocarbon input (Figure 2.16).

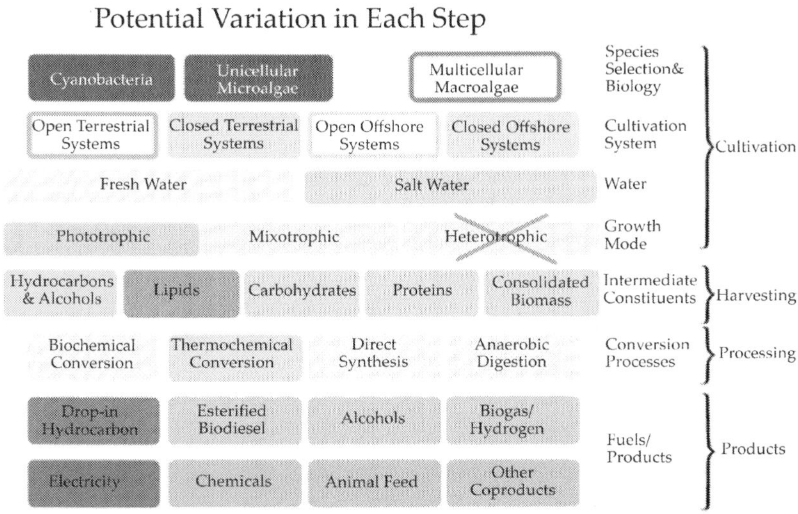

Figure 2.16: Reference pathways: Steps involved to produce green diesel by using an open raceway pond

Beneath the reference track, creating chose lipid and lipid recuperation species are changed over by known substance forms for the generation of hydrocarbons. The concoction structures of these hydrocarbons are without oxygen and are reasonable for use in flying and street as fuel. Recuperation of lipid forms most often depicted requires that cells are pulverized layer (lysate) cells and discharge intracellular rumpidas for oils. Triglyceride algae can be prepared in different ways. In like manner, remaining cell mass may have distinctive employments. Two of the utilizations under thought are reused back to algae development or deal as a co-item. In the above case, biomass is vigorously treated to create biogas for vitality era, with the gushing being come back to culture to give the fundamental supplements. This treatment recuperates all phosphate and nitrogen (N) biomass utilized amid development.

The reference track is additionally opened up in Figure 2.17, which demonstrates the points of interest of the machining stages. For this and different figures, they have received diverse traditions. Yellow precious stones demonstrate the sources of info and yields of the orange framework. The procedures of the green procedure speak to those related with straightforward algal culture gathering with lipids and blue dark with compound procedures. The algae are gathered and broken amid the extraction of the oil, bringing about oil lipids and removed biomass. The reference track accept that algal biomass is recuperated first flocculated by including a compound operator, for example, chitosan. The material is recuperated by separated air buoyancy (DAF). DAF, the air is acquainted with increment surface algae where they can without much of a stretch recoup. DAF expands the level of algal biomass to fluid, however centrifugation is as yet expected to decrease the water content for consequent oil recuperation. The separated lipid is anaerobic biomass processed to create great vitality and supplement content is come back to the algae culture. Hydrotreating lipid algae to create immaculate hydrocarbons are like the creation of second-era biodiesel, in view of seed yields to deliver green diesel. This way speaks to what the commission accepts is a standout amongst the well on the way to deliver algae squander on the premise of further developed current advances to develop and preparing ways.

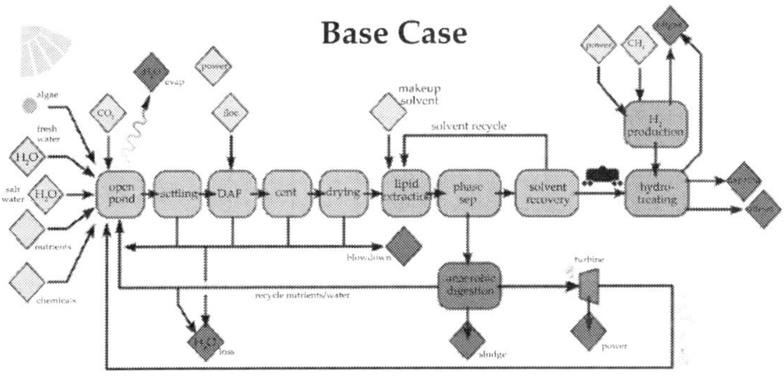

Figure 2.17: Inputs and outputs of the reference pathway.

NOTE: Reference pathway utilizes open raceway lake to deliver algae for preparing to green diesel. Tankcar image reflects just the choice for isolating the lipid generation remote to the fuel handling. DAF is broken up air buoyancy, penny is centrifugation.

Hydrotreatment is utilized to change over crude triacylglycerol hydrocarbon input. Hydrotreatment is a conventional refinery operation that takes out approaching fuel eteroatoms, hydrogenated types of oleaginous alcohols and, conceivably, a chain part. Its fundamental capacity as to lipid is to expel oxygen from the crude material, influencing alkanes to out of the lipid chains, propane out of the principle part of glycerol and water all the oxygen particles. Lipid preparing of algae for hydrotreatment offers a few points of interest. Initial, a fuel substitution happens. Suspended fluid fills are progressively alluring, and its generation is made of triacylglycerol algae gathering as middle of the road. This middle of the road can be proficiently transported to refineries for consideration in the ordinary fuel pool. Triacylglycerol can likewise be dealt with close to the social structure. There are points of interest of scale in hydrotreating transport could support a bigger structure controlled by numerous algae ranches. Among them is the hydrogen necessity. The hydrogen supply pipeline from an incorporated or committed plant presumably has critical monetary advantages. Utilizing current handling assets gives a cost advantage amid generation. Another favorable position is that the heteroatoms notwithstanding oxygen are expelled as in traditional refinery operation. At long last, hydrotreatment is known by the operation of the current unit in refineries. In this manner, joining into existing resources is a moderately smooth progress.

A great part of the vitality talk in the past segment manages expanding vitality utilization and segment collecting, which are encoded in green in Figure 2.17. Numerous LCA considers on vitality utilization and ozone harming substance emanations (GHG) that did exclude the up and coming era of green diesel creation in their investigation. The mode delineated in Figure 2.17 presents fossil vitality usage as hydrogen generation for use in refinery operations. LCA on hydrotreating algae biofuels are not yet accessible. Subsequently, perceptions on vitality utilization and LCA this reference track are made on the premise of analogies. Examination investigations of hydrocarbons entering ordinary generation of esterified biodiesel (FAME) recommend that transformation could be comparative or better to traditional biodiesel as far as petroleum product sources of info and ozone depleting substance discharges. Thus, it can be concluded that reliant hydrotreating generation strategies can possibly be comparable to or superior to traditional biodiesel by furnishing a fuel with preferable properties over FAME.

The reference track demonstrates a dry procedure in which the biomass is gathered and dried before oil extraction. In this procedure, oil extraction is done with the guide of solvents requiring cleaning before reuse. Figure 2.18 demonstrates the vitality and water prerequisites for a way with that dry procedure. Reaping, drying and evacuating parts require critical levels of vitality input and have the potential for advancement to diminish worldwide vitality utilization. Wet procedures where the cell layers are hindered in the fluid medium to discharge the lipids, isolating them to consider gathering, the stage may have a fundamentally bring down drying power utilization. Phosphorus (P) inside the body as phospholipids displays a more noteworthy multifaceted nature in fuel algae preparing. These atoms are unsafe for downstream handling and end utilize can repress or harm impetuses (increment response rate) utilized as a part of fuel change and exhaust systems can harm the vehicle. ASTM D6751-11b, which is the standard detail for biodiesel capacity blending for center distillates, indicates the most extreme approved P content in biodiesel stockroom. P are expelled utilizing fixing innovation produced for use with seed oils. All the more generally, phospholipids end up noticeably immiscible solids which are then expelled by centrifugation. The phospholipid extraction techniques incorporate water, corrosive or enzymatic cleavage. The most regularly utilized corrosive is phosphoric corrosive. The two techniques create a rough elastic stream for transfer or option utilize. Corrosive and enzymatic biting, while it is really reactant, requires corrosive expansion and catalysts.

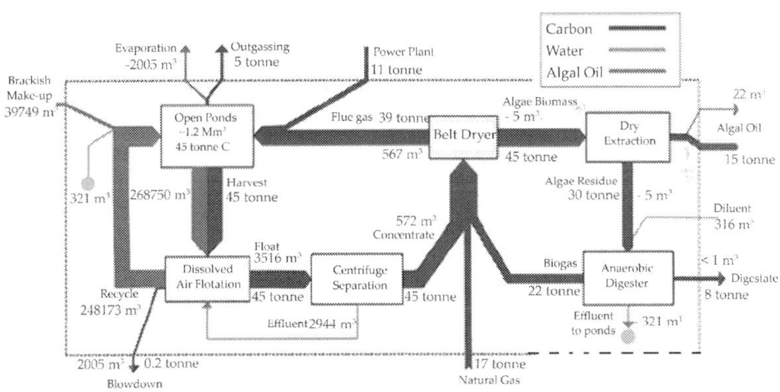

Figure 2.18: Flow of water and carbon in an open raceway pond to produce green diesel.

Raceway Pond Producing Drop- In Hydrocarbon and Co-products (1st Alternative Pathway):

The next line, shown in Figures 2.19 and 2.20, assumes that algal biomass is sufficient to provide a significant revenue stream value. This path involves biomass having original and dehydrated status, and subsequent processing is performed to recover precious smaller components. This pathway translates into an increase in nutrient requirements for algae growth compared to the reference track due to the loss of biomass nutrients from direct co-product values.

Biofuel generation hugy affects volume, so esteem is co-created. The Committee trusts that co-generation of high-esteem items, for example, concoction crude materials, biofuels would be conceivable just on a little scale. On the off chance that vast amounts of high-esteem co-delivered biofuel algae, co-created esteem will probably diminish with immersion in the market. An item that can have a sufficiently vast market to retain extensive amounts delivered is the bolstering of creatures. The co-creation esteem relies upon the structure of the encourage and the attributes of the market where it is sold.

It has been proposed that high esteem items and biofuel co-creation algae is a procedure to address the test of financially attainable algae biofuels. The methodology has been disputable on a few levels. Spreads are firmly connected to the economy and ACV of biofuel generation algae. Spreads are proposed as a way to enhance the economy requires the generation of algae biofuels. Be that as it may, the financial advantage is taken a toll and a basic examination is exhibited to clarify effects and potential concerns. Initially, some broad remarks can be made on the premise of distributed articles and introductions to the advisory group. Maybe the most grounded articulation heard methodology including Fuels, Inc. in a declaration before the council. The moderator trusted the economy for the creation of biofuel algae at a cost that is focused with petroleum derivatives is unthinkable. The concentration of the organization is moving to the creation of co-delivered angle oil yield. The vast majority of the wage is currently originating from fish, and every one of the oils delivered (angle oil for dietary supplements for people or creatures and biofuels for transport) are sold. Others have received a more direct approach and contend that co-items could add to the reasonability of algae biofuels as the market creates and the cost of delivering algae biofuels diminishes with changes and economies of scale proficiency.

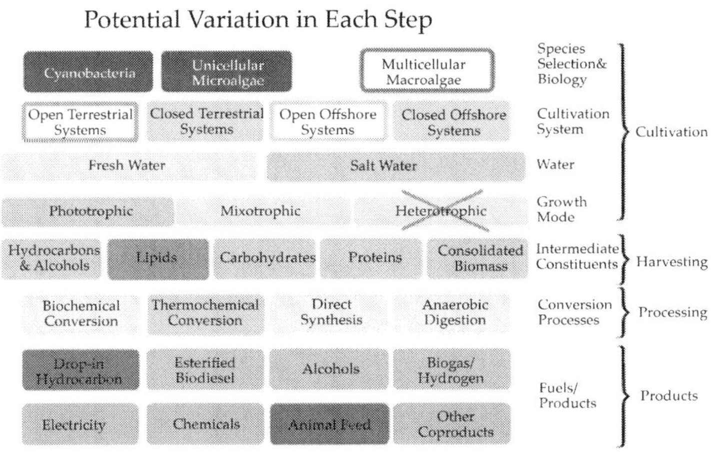

Figure 2.19: 1st Alternative pathways: Production of green diesel in an open raceway pond

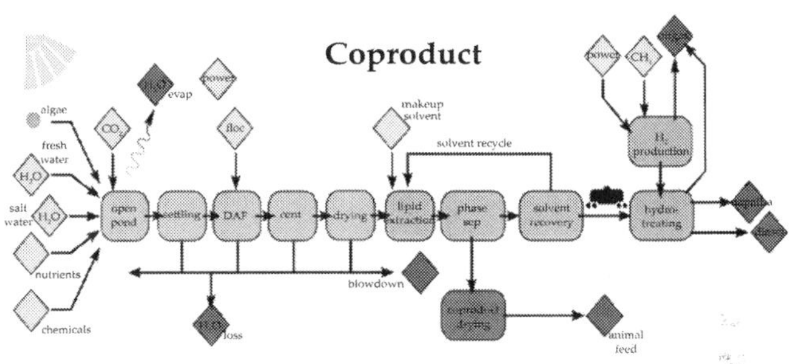

Figure 2.20: 1st alternative pathway: Inputs and outputs during the production of green diesel with animal and fuel feedstuff as co-products in an open raceway pond.

NOTE: As scale expands, cost by and large abatements. This is genuine both for fuel segments and coproducts. The dabbed line demonstrates Szmant's unique bend and the strong line is swelling rectified to 2010.

The measure of biomass increments quickly by expanding the size of biofuels. Remaining biomass is an element of the measure of fuel created and the portion of aggregate dry biomass. Following the lingering biomass as an element of the lipid portion of yearly creation and fuel demonstrates the size of the issue (Figure 2.21). A refinery biofuel alga is comparable to a normal trans-esterification 95 million liter (25 million gallons) biodiesel refinery creates more than 180,000 tons of waste biomass in a small amount of 30 percent lipid estimate. In the event that a division fit for providing 3.8 billion liters (1 billion gallons), which is still just a tenth of that of ethanol fuel in the present United States, a biomass buildup of 7.7 million tons was fabricated.

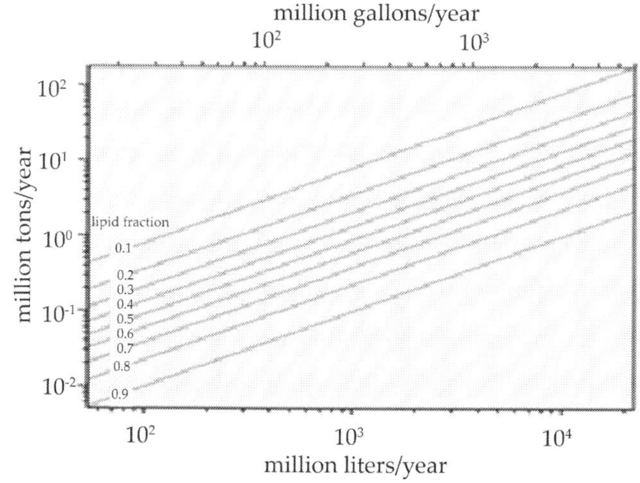

Figure 2.21: Production of residual biomass per annum

NOTE: A solitary algal biofuel refinery is probably going to have a limit of 90 million liters for every year. Current algae are developed in the 20 to 30-percent lipid mass portion go. Five percent of yearly U.S. utilization of transportation energizes is around 39 billion liters.

As talked about with reference to the reference track, this waste biomass can be utilized as a part of anaerobic processing to deliver vitality, yet some remain and ooze to be discarded. Elective # 1 pathway considers the utilization of this deposit to create co-items as nourish.

The level of advertising tends to relate the extension and offering value, which is the most financially savvy return on capital. This is regularly ignored, while co-items are advanced as a critical wellspring of extra income for a monetarily presume fuel generation process. There is fairly little connection between's various items, yet for a solitary item, and the value table is connected to a power charge.

This demonstrates multiplying the scale decreases the value dramatically increased. For materials bound to be sold in mass market fuel mass expanded fuel creation quickly co-delivered unless an extensive variety of auxiliary items for various markets happen. From the angle of asset maintainability, the entirely depicted reference track speaks to monetary examination and stroke is finished. The utilization of anaerobic processing for supplements to come back to developing algae and the supply of vitality for the generation of biofuel algae is a key part of option course # 1. Expulsion of lingering biomass as a co-item in this manner hampers generation Fuel vitality adjust and heaps of supplements required.

Raceway Pond Producing FAME (second Alternative Pathway)

Distinction is not a hydrocarbon fuel, but rather an ester made by trans-esterification of triacylglycerol. This way is all the more intently approximated to customary bio-oil biodiesel moves toward becoming fuel-proficient (Figure 2.22). Triacylglycerol algal is responded with methanol to frame FAME or what is called biodiesel (Figure 2.23). Popularity has poor frosty stream properties and can not be utilized as immaculate segments in chilly situations. Expanding thickness with diesel fuel FAME makes it hard to pump. Indeed, even fix cloud issues mixes may happen when wax precious stones start to shape. Wax gems can prompt gel arrangement, which is contrary with motor operation. Blends containing biodiesel have higher cloud focuses and pour focuses (the temperature at which the fuel is gelled for no more stream) of immaculate diesel hydrocarbons. Therefore, biodiesel is by and large blended with oil-based diesel for definite utilize. Because of its higher oxygen content, FAME is 10 percent bring down vitality substance of diesel hydrocarbons; consequently, it has diminished the mileage of the utilized vehicle per gallon. Acclaim biodegradable long haul safeguarding because of its compound movement and presentation to air and water quickens corruption. In any case, DMARDs can be made generally effective in little scale for algae handling and last fuel generation can happen on a similar site. Likewise has low sulfur and fragrant mixes, and along these lines low particulate emanations when the fuel is scorched. Glycerol co-product is likewise created along these lines; however glycerol has low market esteem.

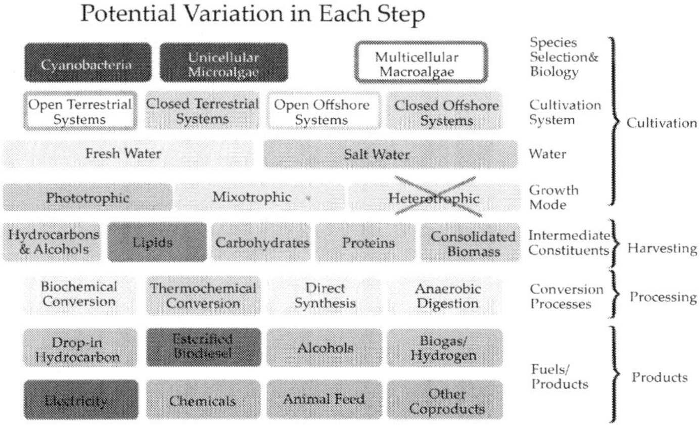

Figure 2.22: 2nd alternative pathway: Production of esterified biodiesel (fatty-acid methyl esters (FAME)) in an open raceway pond

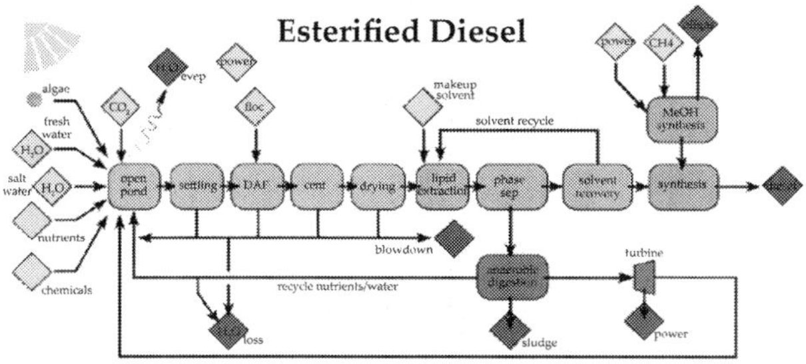

Figure 2.23: 2nd alternative pathway: Inputs and outputs during the production of fatty-acid methyl ester (FAME) in an open raceway pond

Photobioreactors with Direct synthesis of Ethanol (3rd Alternative Pathway)

The procedures depicted above for the creation of algae biofuels are focused on open frameworks for algae lake development, and most breaks down show that frameworks are restrictive photobookers for the cost of delivering fuel. At present, photobioreactor frameworks are utilized to deliver algal biomass of high-esteem items, for example, nutraceuticals and restorative fixings. The course portrayed underneath accept that a marine algae or cyanobacteria species specifically creates a valuable fuel item (Figure 2.25).

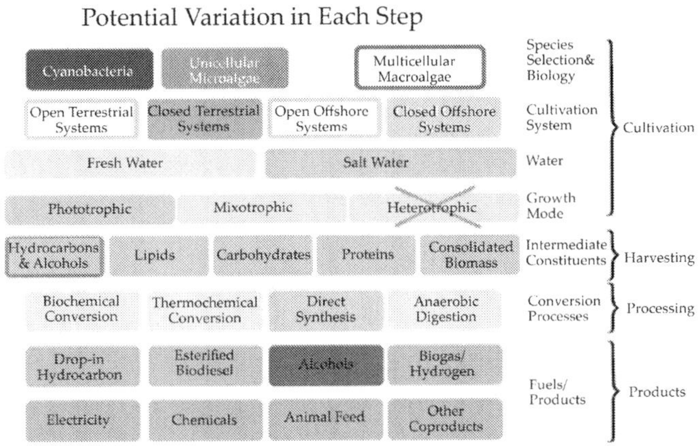

Figure 2.24: Direct synthesis of algenol by using closed reactor

Coordinate union of fuel parts requires for all intents and purposes no algae cyanobacteria are developed in a shut photobioreactor to anticipate debasement of the item. This choice varies impressively from the creation of lipids that require cell film break to gather. The gathering is nonstop with the body discharging the item into the media constantly. Development of algae or cyanobacteria can be steady for quite a long time. This way cannot be done in an open lake since it is trusted that the blend rate of fuel is about the same as the rate of microbial debasement. What's more, fowls will be lost because of dissipation (Figure 2.24 and 2.25).

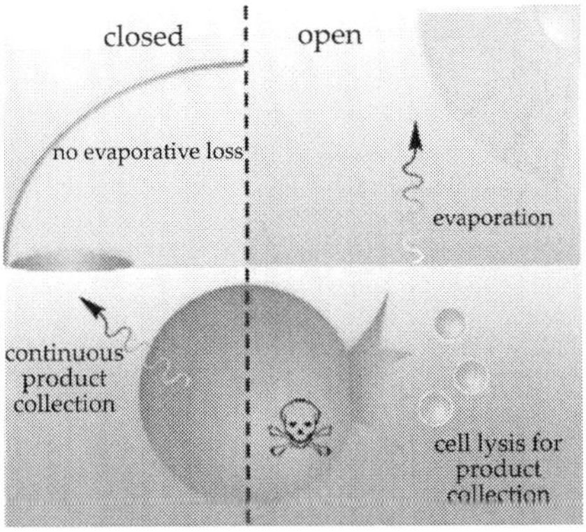

Figure 2.25: A graphical representation to highlight the difference between direct synthesis and the cell-lysis for collection of lipids

A solitary algal biofuel refinery is probably going to have a limit of 90 million liters for every year. Current algae are developed in the 20 to 30-percent lipid mass portion go. Five percent of yearly U.S. utilization of transportation energizes is around 39 billion liters.

Algenol reactors are polyethylene bags. Cyanobacteria release ethanol in the vehicle carrying, which is then divided between the liquid space and the photobioreactor head. The ethanol produced is trapped in the closed photobioreactor. The solar energy that penetrates the bag forms a sort of «solar still». Primary diluted ethanol collection is essentially determined by solar energy, but the subsequent purification phases require the input of energy derived from fossil fuels to produce specific alcohol fuels encounter

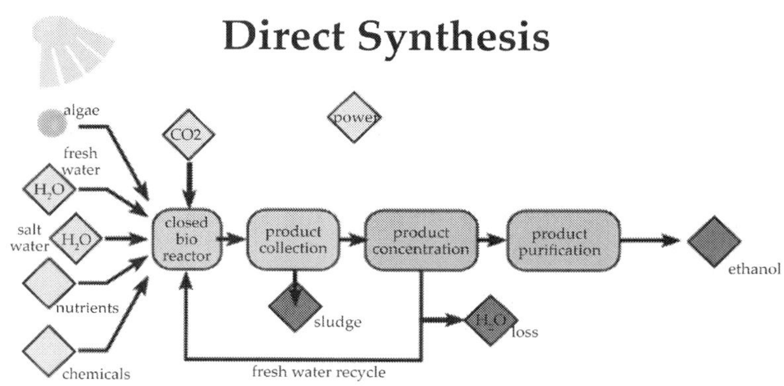

Figure 2.26: 3rd alternative pathway to represent the inputs and outputs in a Photobioreactors for the direct synthesis of ethanol

Algenol measured the focus in the condensate in the ethanol photoboreactor and observed to be 0.5 and 2.0 percent. The vitality required to recuperate the usable fuel liquor as an expansive part decides the vitality productivity and, in this manner, emanations all the while. The outcomes are for the most part great vitality adjust; That is, more vitality is held in the fuel that is expected to do as such. The ROI vitality this way motioned by an Algenol agent board of trustees is inside the 1.2 territory generally detailed for corn ethanol and sugar stick 8. Be that as it may, these outcomes and those of other biofuel algae frameworks have not been gotten from plants completely illustrated.

Examination of results with other Algenol algae biofuel ponders is great regarding other vitality asset necessities and. Wiping out the requirement for drying out decreases vitality necessities and is a reasonable favorable position of procedures that straightforwardly create fuel. They are additionally natural to a photobioreactor framework shut the advantages of diminishing water and supplement utilization and decreasing the danger of sullying. The cost of capital is a generally specified worry for shut bio-filled photocatalyst frameworks, yet not tended to in this report.

Water utilize can be fundamentally lessened in a shut bioreactor yet can not be totally disposed of. Water is lost in photosynthesis and preparing. On the off chance that marine or freshwater algae are utilized, the expansion of new water or crisp water is important to keep up the water level and key fixations, for example, saltiness. Photosynthesis devours water:

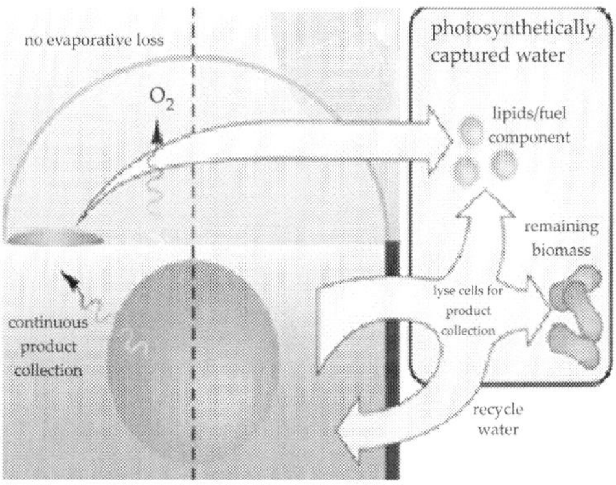

Figure 2.27: Consumption of water during photosynthesis for the production of algal biomass for fuels

For each settled carbon, no less than one particle of water (condition) is responded. In spite of the fact that there is no spillage misfortune and process water is totally reused, water is devoured because of photosynthesis. This sum can be assessed by the stoichiometry of the photosynthesis and the creation fuel. Cutoff cases for a shut framework are appeared in Figure 2.27. Water expended amid photosynthesis relies upon the circulation of carbon settled in the biomass or fuel part. For the Algenol configuration said thus, every one of the carbons are really settled in the fuel part which is ethanol having a nuclear weight of 46 grams g/mol. The normal atomic weight is 46/2 or carbon 23 grams/mole. The normal sub-atomic recipe for carbon biomass is CH_2O, with a normal sub-atomic weight of 30 grams of carbon for every mol. Lipids have a normal sub-atomic weight of around 16 grams of carbon for every mol.

Water devoured amid photosynthesis relies upon the appropriation of carbon settled in the biomass or fuel segment. For the Algenol configuration said thus, every one of the carbons are really settled in the fuel part which is ethanol having a nuclear weight of 46 grams g/mol. The normal atomic weight is 46/2 or carbon 23 grams/mole. The normal atomic equation for carbon biomass is CH2O; with a normal sub-atomic weight of 30 grams of carbon for each mol. Lipids have a normal sub-atomic weight of around 16 grams of carbon for each mol.

Entire Cell Processing (fourth Alternative Pathway)

Thermochemical approaches to process fuel biomass pulled in enthusiasm as the concentration moved from liquor and remote creation and drop-in hydrocarbon generation. Figure 2.29 demonstrates the passage and leave points of interest of a pathway of thermo-substance pyrolysis. Anaerobic for all intents and purposes any biomass warming causes warm corruption starts to fractionate biomass into vaporous and strong, fluid segments. In this manner, fluid portions can be enhanced hydrotreatment to deliver a hydrocarbon fuel. Last fuel items are good with the current oil refinery framework

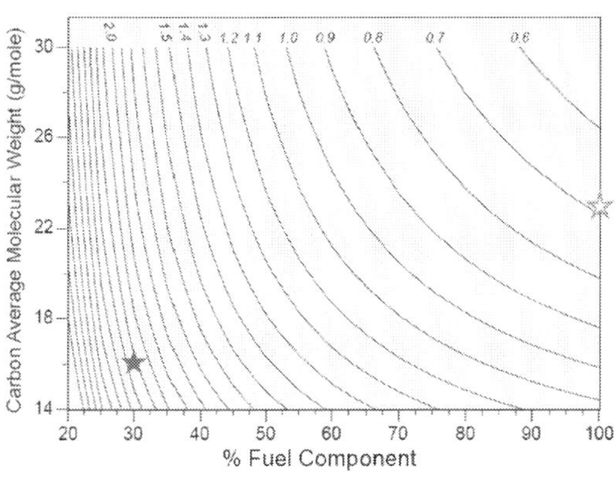

Figure 2.28: Graph of water (volume ratio) consumed per liter in photosynthesis

NOTES: The orange star approximates the Algenol coordinate ethanol amalgamation accepting that ethanol is delivered in a steady culture that creates no leftover biomass. The red star is appeared for reference and approximates the case for a lipid shaping algae at 30-percent lipid.

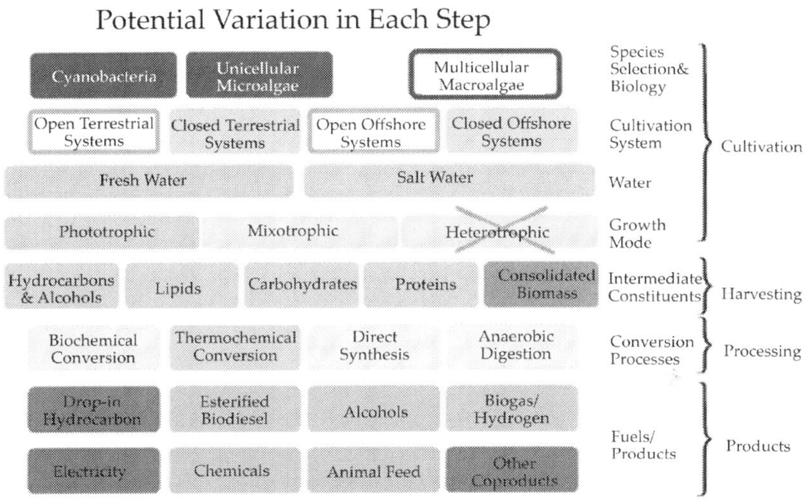

Figure 2.29: Cell processing: An overview

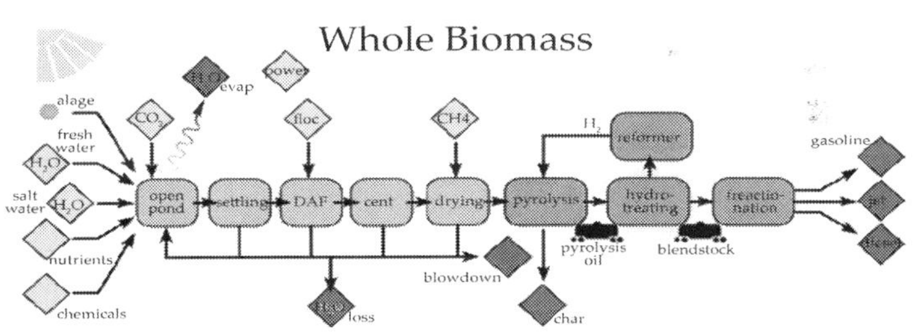

Figure 2.30: 4th alternative pathway: Inputs and outputs during the pyrolysis of cellular biomass and its hydro-treating to yield hydrocarbon blend-stock

NOTE: A few choices exist for the pyrolysis of algal biomass. Appeared here is a two-organize change where tank cars demonstrate that it is possible to transport both pyrolysis oil and blendstock.

Algae must be gathered, dried and most likely dried for use as crude material. Oceanic species shows a test to a pyrolysis procedure because of lead water to the procedure. Water fills in as receptive diluents to a great extent maintaining a strategic distance from warm amid the pyrolysis stage. Diluents, water lessen the productivity of endeavors to recuperate the estimation of gas streams.

There are a few choices for the pyrolysis procedure. Middle pyrolysis oil represent some specialized difficulties. Pyrolysis oil is acidic and responsive, and capacity without adjustment is a test. Hydropyrolysis is incorporated hydro-transformation (IH2) is a hydrolysis and hydrotreating process created by the Institute of Gas Technology and associates and financed by the Department of Energy. IH2 has many promising credits and has asserted to process elite algae appropriately. This specific mix is to keep away from pyrolysis oil issue.

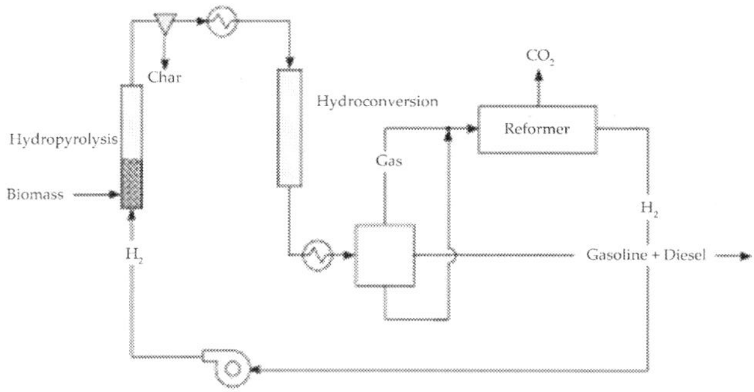

Figure 2.31: IH_2 process schematic.

SOURCE:https://www.researchgate.net/figure/230083622_fig13_Figure-1-IH2R-system-schematic-showing-overall-process-flow

IH2 innovation utilizes low weight hydrogen combined with a property to dispose of for all intents and purposes all present in the biomass beginning (Figure 2.31) oxygen impetus. Exportable steam creation is conceivable and is likely adequate to incompletely balance the vitality interest for drying. The hydrogen required for the procedure is delivered by steam changing of the fumes gas stream of methane and different hydrocarbons from the procedure. Pilot-scale tests have been promising and various exhibition units under development. It demonstrates that the potential for ozone harming substance decrease is lower than contending forms. The writing on the execution of this procedure is restricted. Burning of fuel items extended from 26 to 46 percent on a dry, ashless premise, contingent upon the crude material. This speaks to the high vitality productivity change of more than 70 percent for most extreme execution. This procedure was decided for examination because of reports proposing that extra vitality and hydrogen are not required. Thus, there are no additional or gushing supply streams that influence the investigation of world efficiency affect generation. The ventures are continuing with this innovation to change over algal biomass.

Other Potential Pathway

Numerous other microalga fuel extraction forms are examined and considered. The absence of distributed or accessible information on key maintainability measurements implies that little can be said in regards to the supportability traits of other potential courses in connection to the modes talked about above. Notwithstanding the pyrolysis way depicted above, begin microalgae frameworks tried utilizing thermochemical change procedures to process finish algae cells. These frameworks appear to have vitality necessities like other generation strategies where cells are executed to gather the item. Unmistakably, thermo-compound pathways can deliver various flammable items. It is misty whether the ensuing utilization of vitality in handling offers vitality proficient advantages and emanations. These full biomass frameworks offer the favorable position that collect is not restricted to oilseeds. The species can develop at greatest carbon settling rates at control forms that keep up high carbon parts settled in their last fuel. Cases of entire transformation advances are:

- Fermentation of algal biomass to yield alcohols or hydrocarbons
- Gasification and syngas transformation to alkanes, alcohols, or aromatics (through methanol and resulting change)
- Traditional catalysis for gasification and syngas transformation to alcohols by
- Gasification and syngas transformation to alcohols by syngas aging
- Anaerobic assimilation to methane (making no fluid fuel)
- Hydrous pyrolysis.

These methods are not exceptionally broad and can be more extensive practically speaking. High water content is not an alluring element of the charged fuel. The rudiments of water evacuation item are crucial in any vast scale fuel creation discourse. Research facility specialists, on a pilot scale utilizing sun based drying, are generally moderate, require expansive regions of land and are not prone to create available

CHAPTER 3

RESOURCES OF ALGAL BIOFUELS

ALGAL BIOFUELS

Algae are the fast developing plants on the planet. Modern reactors for algal societies are of three sorts which are following:

- Open lakes
- Photobioreactors
- Closed frameworks.

Algae are imperative as biomass sources. Presently algae are considered as a wellspring of biofuels. Various types of algae offer diverse sorts of fills (biofuels). Algae don't require a good situation, can be effortlessly developed anyplace, even in waste or salt water, and don't require prolific products or rich sustenance, and preparing requires less vitality than algae. In light of the developing populace of ocean growth fuel oil is utilized as a central reason. Algae can be a substitute for oil powers, which is more compelling and has inconveniences. As we probably am aware, algae are among the quickest developing plants on the planet, and around half of its weight is the oil called lipid oil. This oil lipid can be utilized to deliver biodiesel (car fuel) for autos, Lorries and planes. Microalgae have rates substantially speedier than developing harvests on Earth. It is assessed that the yield per hectare of ocean growth oil is from 20,000 to 80,000 l for every section of land every year; this is 7-31 times higher than the following best palm oil collect. The substance of lipids and unsaturated fats from microalgae shifts relying upon the way of life conditions. The greater part of the ebb and flow oil extraction investigate concentrates on microalgae for the creation of ocean growth biodiesel. Oil algae forms in biodiesel effectively as oil got from arrive crops.

Fuel algae are not quite the same as petroleum product utilize algae as a wellspring of normal stores. A few organizations and government offices are financing endeavors to lessen working and capital expenses and make monetarily reasonable fuel creation of algae. Seedlings, for example, non-renewable energy sources, discharge carbon dioxide when copied, yet not at all like petroleum derivatives, carbon dioxide is expelled from the air by developing algae and different wellsprings of biofuels. The vitality emergency and the worldwide sustenance emergency have powered the enthusiasm of algiculture to deliver vegetable oil, biodiesel, bioethanol, bio-fuel, biomethanol, biobutanol and different biofuels by utilizing unacceptable farming area.

CHOOSING THE BEST ALGAE FOR BIOFUEL

Algae are being viewed as the conceivable biofuel without bounds, since a few species yield up to 60% of their biomass in lipids, otherwise known as oil. To place that in context:

- Corn yields around 18 gallons of oil for each section of land.
- Sunflowers yield 102 gallons
- Micro algae yield 5000-15,000 gallons.

There are a huge amount of various sorts of algae on our planet yet some are significantly more suited for biofuel than others. You need to discover one that is anything but difficult to develop, produces a high level of oil, and duplicates moderately rapidly.

Chlorella

Chlorella is the sort of algae. It is genuinely prevalent among cultivators since it has a quick development rate and develops well in a lake setting. Around 29% of its dry weight is oil, in light of the fact that distinctive sorts of algae develop in various atmospheres, ought to pick the algae which the developing rate is exceptionally fast and give best yield. Algae are rising to be a standout amongst the most encouraging long haul, supportable wellsprings of biomass and oils for fuel, nourishment, sustain, and other co-items. These are so alluring, the extensive number and wide assortment of advantages related with how and where they develop.

Figure 3.1: Chlorella

SOURCE: http://www.lukeshavak.com/miessence/chlorella-benefits/

Chlorella stores vitality as fuel. About every one of these advantages originate from the way that these plants have developed more than billions of years to create and store vitality as oil, and they do this more proficiently than whatever other known characteristic or designed process(extract the oil for the employments of oil-based purposes).

Algae are indistinguishable whatever other photoautotrophic living being. They turn basic inorganic mixes, for example, CO2 and water into complcx, vitality rich hydrocarbons utilizing light as a vitality source. As we probably am aware the significant part of oil is Hydrocarbons. The innovation and framework to make fuel from this kind of biomass have existed for no less than 60 years.

Plant matter (lignocelluloses) is the most bounteous crude material on the planet for the creation of ethanol by maturation, and numerous biofuels are made utilizing the oils from harvests, for example, corn or soya beans. Presently the uprooting impact of utilizing land crops for fuel creation rather than sustenance seemingly causes more carbon outflows through deforestation and can drive up nourishment costs.

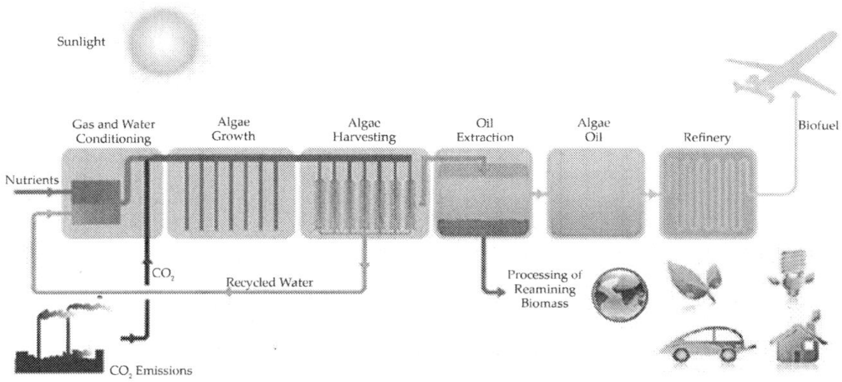

Figure 3.2: Source of biofuel and its uses in the world

SOURCE:http://teambiofuels.weebly.com/generation-iii-biofuel.html

Algae created a considerable measure of oils in a brief timeframe when cells approach sun based vitality. For the creation of ocean growth oil is vital and capacities like put away vitality and access to light. Kelp produces gravity at the highest point of the water section on the grounds that the lighter oil of the water. Algae have no characteristic drive, so oil generation enables access to more photons that bolster photosynthesis. In regular habitats, ordinarily the algae species in the best have a large portion of the oil. Algae give distinctive sorts of fills of various types of algae.

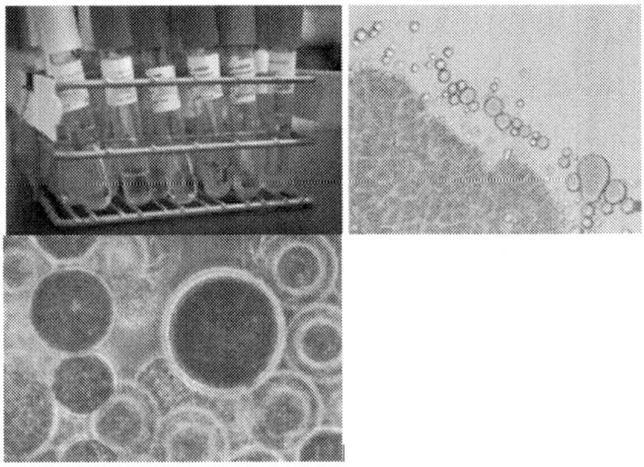

Figure: 3.3: Bioprospecting for Oleaginous Algae

SOURCE: http://www.21stcentech.com/wp-content/uploads/2012/04/N-gaditana.jpg

Recent technologies introduced allowing manufacturers to scan thousands of algae cells to find those that produce robust and fast, while at the same time offering high oil content, often about 40%. Some manufacturers are stressed algae by maintaining a nutrient like nitrogen, which causes algae to protect the environment and also protect oil overproduction.

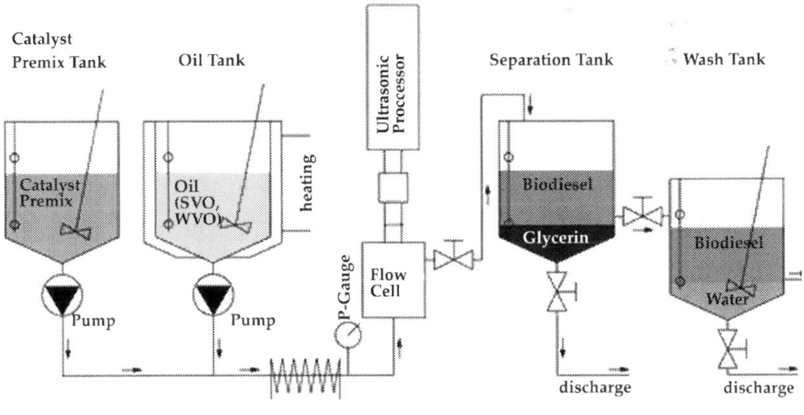

Figure: 3.4: Productions of Biofuels (Biodiesel)

SOURCE: https://www.intechopen.com/books/biofuels-status-and-perspective/technologies-for-biodiesel-production-in-sub-saharan-african-countries

Algae Biofuel Production

Biogas gathered algae experiences the lysis procedure that exchanges and enables the oil to buoy to the surface of the fluid. Lisi can be created from solvents, for example, hexane, chemicals, electrical weight, mechanical (printing) or laser. It goes about as a supercritical carbon dioxide when pressurized and warmed to change its sythesis from fluid gas. Supercritical CO2 blended with algae separated just about 100% of oil, however the procedure is more costly than other oil extraction strategies. Late research demonstrates that algae biofuels are more advantageous than different wellsprings of oil. Oil makers culminate the objective biofuels, bioplastics or fine green chemicals. Refining of oil is utilized as a part of transesterification of unsaturated fat chains. A liquor, for example, methanol and an outside compound are blended to make a response to deliver an alternate sort of liquor and esters. This procedure is additionally used to create polyester textures. Esters are substance mixes having their corrosive hydroxy gatherings, an atom of hydrogen and oxygen together, supplanted by a particle of oxygen. The substance transesterification response progresses toward becoming ocean growth oil in biodiesel. These biodiesel (biofuels) are utilized as a part of the vehicle procedure.

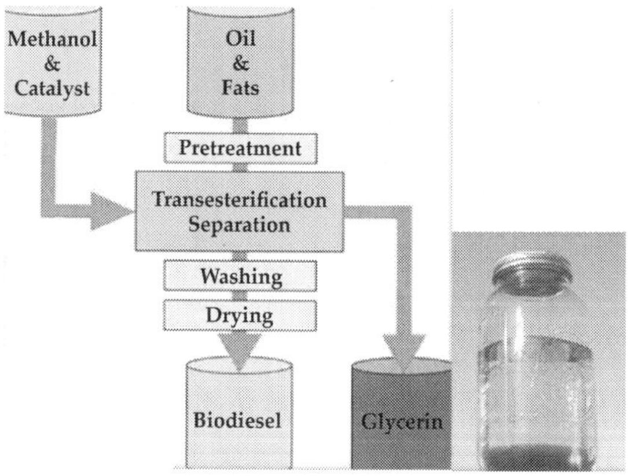

Figure: 3.5: Refining Biofuels from Algae Oil

SOURCE:https://www.slideshare.net/madhujegathish/microbial-biodiesel-production

Lingering biomass, most proteins and sugars, can be utilized for nourishment, encourage, composts or supplements, numerous different items. Huge scale creation of algae biofuels produces generous advances for creatures and plants. For the most part, biofuel protein coproducts are not sufficiently perfect for human nourishment utilize. In any case, one of the most grounded advantages of biofuel generation is the expansion in R and D exercises that will profit all algae makers.

Algal Nutrient Solution

Supplements, for example, nitrogen (N), phosphorus (P) and potassium (K) are critical for plant development and are basic parts of manure. Silica and press, and different follow components can be viewed as imperative marine supplements; in view of absence of one can restrict development or efficiency in a zone.

Carbon dioxide

Gurgling CO2 through algae development frameworks can extraordinarily expand profitability and effectiveness (up to the immersion point). Commonly, around 1.8 tons of CO2 for every ton of algae (dry) biomass created, however this fluctuates with the algae species. Glenturret Distillery in Perthshire, UK - home to the renowned grape bourbon - permeated CO2 made amid refining of bourbon through a microalgae bioreactor. Each ton of microalga ingests two tons of CO2. Scottish Bioenergy, which coordinates the venture, offers microalgae as rich in protein and high-esteem fisheries. Later on, it will utilize algae deposits for sustainable power source creation through anaerobic absorption.

Nitrogen

Nitrogen is an imperative substrate that can be utilized as a part of algae development. Different wellsprings of nitrogen can be utilized as algae supplements with various limits. Nitrate of nitrogen source, contrasted and the measure of grown-up biomass is favored. Urea is likewise accessible for the source that shows equivalent outcomes, so it is a sparing substitute for the nitrogen source in vast scale algae development. In spite of the unmistakable increment in development over the nitrogen free medium, this demonstrates modifications in nitrogen levels influence the lipid content inside the algae cells. In a 72-hour nitrogen hardship, add up to unsaturated fats (in a cell base) expanded 2.4 times. 65% of the aggregate unsaturated fat esterified to triacylglycerols in sleek bodies contrasted with the underlying society, demonstrating that algae cells utilized novel amalgamation of unsaturated fats. It is critical that the lipid content in the algal cells is adequately high, while the proper cell division times are kept up, so the parameters can boost both are under thought.

Wastewater treatment facility

The conceivable wellsprings of supplements are sewage, water or farming water, which are as of now the principle toxins and wellbeing dangers. Notwithstanding, these effluents cannot encourage straightforwardly algae and should first be handled by microbes, through anaerobic assimilation. On the off chance that untreated sewage before they achieve the algae, algae sully the reactor and in any event slaughter a large number of the algae strains required. In biogas plants, natural waste frequently turns into a blend of carbon dioxide, methane and natural composts. Fertilizer is utilized leaves digestor is fluid and reasonable for algae development yet first it must be cleaned and sanitized.

Figure: 3.6: Algae Farms & Artificial Rainforests

SOURCE:https://www.greentechmedia.com/articles/read/petroalgae-a-zero-revenue-company-files-for-ipo

The utilization of seawater and sewage, rather than crisp water, is unequivocally protected because of the nonstop consumption of freshwater assets. In any case, substantial metals, metal follows, and different contaminations in wastewater can lessen the capacity of cells to deliver biosynthesis lipids and furthermore influence numerous different work environments in cell systems. This is valid for sea water, yet contaminants are at various fixations. Hence, rural review wellspring of supplement is favored, however overwhelming metals are again making an issue, particularly for algal strains that are powerless to these metals. In open tank frameworks, the utilization of algal strains that can deal with high centralizations of substantial metals could keep different living beings from damaging these frameworks. At times it has likewise been demonstrated that algae can expel over 90% nickel and zinc mechanical wastewater in moderately brief time.

ALGAE FARMS & ARTIFICIAL RAINFORESTS

Albeit Green and Hydra are intended for singular properties, there is an incentive in biomass generation plants on a modern scale (and potentially vitality). Develop Alghe (Energy) can outline biomass generation places for governments and organizations to add to the accessibility of assets for free regular citizen and military focuses, research or gear that could be disconnected.

Likewise, association with Industries, Grow Energy is building up a compassionate arranged model for what can be depicted as "simulated rainforests" substantial, mechanized biomass generation offices intended for the creating scene, which impersonate the carbon-engrossing advantages of normal rainforests.

Algal microfarms the future of urban farming

Developing nearby sustenance in group gardens and urban ranches is a well-known and developing pattern. Since urban space is restricted and valuable, a few business people are transforming old distribution centers and underground spaces into indoor ranches with manufactured lighting. Others propose vertical homesteads, additionally developing sustenance in high rises. Yet, a few researchers evaluate that indoor urban cultivating brings about nourishment generation at a high cost, high vitality utilize, and a substantial carbon impression, paying little respect to whether the vitality originates from customary or sustainable nets. They caution that developed sustenance with simulated indoor lighting can be maintainable and less unsafe to nature than nourishment transported by 3000 miles away. Growing nursery sustenance in free parcels or close urban ranges would be a less expensive arrangement. Urban nurseries would expand the development season, utilize existing urban foundation, appreciate access to nearby and client showcases and grab the best sun based vitality source.

In the course of the most recent 30 years, algae spp. (Spirulina) has won overall acknowledgment as a superfood supplement. Universally distributed research indicates little measures of day by day increment safe capacity, enhancing detoxification, reestablishing gainful intestinal vegetation, enhancing neurologic capacity and knowing the body's recuperating reaction. Spirulina controls the retail value premium, $ 80/ lb for the dry item in the US, 150 €/kg in France.

The vast majority consider huge algae ranches situated in the hot betray sun or tropical spots. Be that as it may, in the course of the most recent decade, more than 110 micrograms of nursery algae have spread all through France. In the United States, Spirulina is effectively developed in microfármacos nurseries in the northwestern North Pacific atmosphere.

Figure 3.7: Retractable and removable greenhouse covers for Algae growing ponds to adjust the temperature in different seasons of the year

SOURCE: http://www.smartmicrofarms.com/category/blogs/

These new algae micrograms are particular and versatile frameworks for insightful nursery checking and web cameras, so remote specialists can effectively drive neighborhood operations in many spots at one and a similar time. With confined lakes and effective LED lighting and warming, the period of algae development can stretch out into Nordic atmospheres past the common 4 or 5 months. Through this system the development of algae is more.

RESOURCES OF ALGAL BIOFUELS

Fuel generation from fossil and organic crude materials numerous assets and algae biofuels require asset data sources, for example, water, vitality, soil and supplements. Algae biofuels have been delivered on a little scale, enough to demonstrate that there are various conceivable methods for generation. In spite of the fact that the generation of algae biofuels is in fact possible, they have turned out to be monetary, ecological and social to end up plainly a feasible substitute for oil energizes. Acceleration courses for the creation of algae biofuels which are viewed as down to earth for business generation postures new requests on regular assets. Watchful thought ought to be given to levels of supplements, water, soil and vitality for the generation of option, monetarily, naturally and socially reasonable option biofuels. This part concentrates on ebb and flow learning of the supportability of water, supplements, soil and vitality algae biofuels at all phases of the photoautotrophic algae creation process (Figure 3.8). Moreover, potential rating markers are given when there are holes in fitting and current information and information for each of the four classifications of asset needs are distinguished.

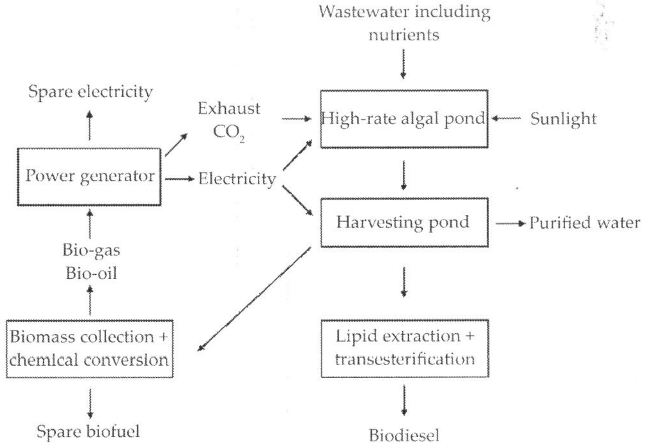

Figure: 3.8: Resource requirements of algal biodiesel production. SOURCE:https://www.researchgate.net/publication/44901352_The_Potential_of_Sustainable_Algal_Biofuel_Production_Using_Wastewater_Resources

New real advances in the present information construct will require exhibit offices in light of numerous hectares to be manufactured and kept up in operation for an adequate timeframe to permit inside and out investigation of the key factors required for business achievement. What's more, the important business scale exhibitions will be to assess and enhance the advances of algal biofuels and their incorporation with existing vitality framework. The consequences of diminished stock usage of assets along the whole algae store network biofuel, dispensing with a portion of the current organizations to grow expansive scale biofuel algae, reasonable and monetarily feasible hindrances. Likewise, enhancing biofuel efficiency and algae execution will help decrease asset prerequisites per unit created biofuel.

Water

Water gives the basic physical condition in which the algae developed develop and replicate. It additionally goes about as a warm controller and gives a way to basic supplements:

- Carbon dioxide (CO2)
- Nitrogen (N)
- Phosphorus (P)
- And different supplements for algal biomass generation.

Water ought to be pumped and contained and circulated in mass culture frameworks, including open or shut tank bioreactors. Shut photobioreactors can likewise utilize water shower or jump to keep up the way of life temperature. As the interest for rural water in the United States represents 85 percent of water utilization, huge scale biomass creation, specifically algae, has the potential for provincial weight on water frameworks, unless it utilizes wellsprings of water Fresh water where conceivable.

Despite the kind of fuel delivered, water is a vital piece of the fuel generation, so there is an essential connection amongst fuel and water creation. For the creation of kelp biofuels, water is required for the generation of biomass crude materials, and can be lost amid the handling of algal biomass energizes. This segment shows the prerequisites of water and crisp water enhance amid different phases of the store network algae biofuels and for the duration of the life cycle of algae biofuel creation.

In this report, the water prerequisite alludes to the measure of water required for the duration of the life cycle of algae generation of biofuels. Utilization of new water is the measure of new water removed from the surface or underground springs of water lost in the prompt region by vanishing or consolidation into items. For instance, water removed from a fossil digicle is quickly lessening is less practical than a similar measure of water extricated from a pipe that is immediately provided. At the point when information is accessible, appraisals of water utilization contrasted with different biofuels and fuel oil.

Water Requirements in the Supply Chain

Water prerequisites of any algal culture framework rely upon the physical structure and arrangement of the framework, the nearby atmosphere and the capacity to recuperate and reuse the framework. Open lakes are liable to evaporative water spillages that are impacted by many elements, including the tin territory, volume and water level; Water and air temperature; And wind speed, mugginess and environmental weight. It is assessed that the US normal vanishing rate of a lake framework is 0.9 cubic meters of water per square meter every year, except the dissipation misfortunes of open lakes differ as indicated by the geographic district. Moreover, some working frameworks (eg, tumult and air pockets) can build water misfortune to more elevated amounts than would be normal by dissipation and cleanse as it were. In outside algae development, crisp water is expected to make up for the loss of water by vanishing and avoid salt amassing. Along these lines, new water is required in aalgae culture framework paying little respect to the kind of culture water utilized. The proportion amongst vanishing and draining lose to open in the framework as appeared in Figure 3.8. A lot of water (Fout) is lost by vanishing in open tanks, along these lines thinking the aggregate broke up solids in the pool water. Regardless of whether encouraged as sweet as salted water, all frameworks have an expanding algae control point for the greatest admissible grouping of disintegrated solids being kept up in culture. On the off chance that this settled point depends on saltiness, vanishing would build the saltiness of the pool, so measures ought to be taken to adjust for this expansion. The augmentations of water with a lower grouping of salt and tin wash water are two stages that can be taken to keep up saltiness beneath the characterized control point

(XControl). The two stages can build water request and water utilization. Notwithstanding dissipation misfortunes, water can enter and open tanks, particularly if covered with earth or if a covering disappointment happens. Water permeation is firmly affected by the piece and structure of the hidden soil (eg sand against mud). Penetration rates are ordinarily in the scope of 5 to 6 millimeters for each day, which is low contrasted with dissipation rates in many parts of the United States. Dissimilar to open, shut photoboreactors, which are not influenced by surface vanishing and filtration, and the lower esteems detailed for the utilization of assessed water are related with shut frameworks. In any case, the water necessity of a photoreceptor framework relies upon the genuine setup and operation conditions.Water recovery and reusing are significant for add up to vast water and photovoltaic water framework prerequisites lakes. The sum and measure of culture water can be recovered and reused relies upon the productivity of the detachment forms, water return quality and its own algae affectability that are developing to changes and contaminations from the arrival water from occupant algae.

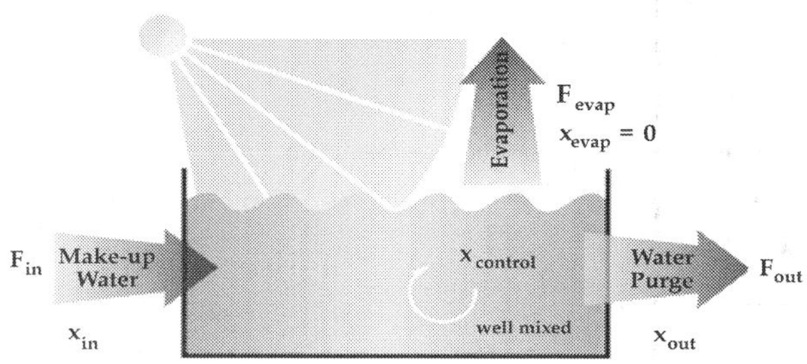

Figure 3.9: Managing solids' (dissolved) concentration in cultivation system of open-pond.

The water prerequisite for the treatment of algal biomass in biofuels is low contrasted with dissipation misfortunes amid development in vast water frameworks. It has been assessed that the utilization of water to process algal biomass of unsaturated fat methyl esters (FAME) was 1 liter for each liter of biodiesel delivered. Loss of water amid drying of algal biomass to get ready for fuel transformation is unavoidable, and definitely water is additionally lost amid seawater oil extraction and oil esterification algae. In any case, Pate et al. (2011) underlined that water spillage under working conditions including the utilization of water with a high salt substance, expanded saltiness to less crisp water is utilized to make up for the misfortune or measures are taken to relieve or Fit-to salt.

Life-Cycle Water Requirement

The measurement of life cycle water interest for algae biofuel creation is oversee future effects on water request and permit examination of the utilization of water for algae biofuels with different energizes. Assessing the lifetime utilization of any prerequisite for the generation of algae biofuels (e.g water, supplements and vitality) is, be that as it may, convoluted by the advancement of innovations. Notwithstanding vulnerability with respect to how biofuel algae advance as far as showcasing, there is likewise an absence of information on the materials and vitality needs of current innovations.

An extra intricacy in the development of the ocean growth is interested in the utilization of water altogether fluctuates with climatic contrasts in temperature, mugginess and precipitation. Different biofuels and products demonstrate this fluctuation in water utilize. For instance, local inconstancy in water system comes about appraisals life cycle water prerequisites to deliver corn ethanol running from 5 to 2140 liters of water for every liter of fuel, contingent upon the state in which corn is developed. In the event that the interest for national normal water for the creation of corn grains is utilized, an exploration built up that water use for corn grain ethanol was 142 liters of water for every liter of fuel. Be that as it may, the land appropriation of additional corn delivered to fulfill the ethanol request is questionable, so it is hazy if its water request matches with the national normal of all corn. The water power of algae lake open products depends basically on the future geological circulation of the yield. This further dissemination is hard to foresee, in any case, on the premise of the vulnerability in innovation execution, arrangement and industry reaction. Without a dependable visualization, water power studies can elucidate the connection between position, atmosphere and water.

Life-Cycle Water Use of Freshwater Open-Pond Systems

A few examinations have broke down the water prerequisites of biofuels created from algae developed in untamed water frameworks and incorporate different life cycle stages. The outcomes incorporate more than two requests of greatness, from 32 to 3650 liters of water for each liter of algae biofuels. By correlation, 1.9 to 6.6 liters of water are devoured to deliver 1 liter of oil in view of fuel from rough bituminous sands or oil. Settling changeability and instability in these outcomes is past the span of the relationship. Rather, the motivation behind the report is to distinguish and organize issues that could influence the long haul manageability of algae biofuels. Characterizing innovative work (R and D) needs in ranges of intrigue could add to the advancement of algal biofuels as a practical piece of the vitality future.

A specialist investigated the life cycle water request of various option fuel transporters, including ethanol and grass grain maize, soybean biodiesel, biofuels and sun powered power and twist algae with developing algae in frameworks of Open and shut photobioreacter lake. The extent of the broke down procedures incorporates water fused into structures and vehicles. As a rule, the utilization of water vanishing and fuel life cycles rule.

Life-Cycle Water Use in Closed Systems—Photobioreactors

A development alga in photobioreactors can possibly take out water dissipation utilization, which can extraordinarily diminish add up to water request. Be that as it may, information for shut frameworks is considerably shorter than open frameworks. A specialist assessed life cycle water necessity of a photobioreactor for 30-63 liters of crisp water per liter of biodiesel framework, in spite of the fact that this outcome depends on master sentiments and not on the experimental estimation of an utilitarian framework. Algenol process is a shut photobiorector that utilizations ocean water and new water. In its natural effect evaluation for a bio-refinery extend at Fort Meyers, Florida, Algenol gauges that the plant would require 3.6 million liters of ocean water and 210,000 liters of crisp water to deliver 100,000 liters Of algae every year (or 36 liters of water for each liter of ethanol and 2.1 liters of new water per liter of ethanol) (DOE, 2010a). New water utilization is identical to 3.15 liters of new water to deliver each liter of petroleum proportionate fuel. Algenol gauges do exclude water utilize upstream of the contributions of its structures.

Algae Cultivation Using Salt or Brackish Water or Wastewater:

Utilization of tolerant salt kelp species permits the utilization of option water sources, for example, ocean water, salted water and harsh water or coproduced methane oil wells, gaseous petrol and coal. These physiological algae adaptability includes finding algae creation in ranges where there are elective wellsprings of water can diminish the utilization of new water in the product. Expanding algae indoor lakes could lessen the attack capability of lakes for undesirable freshwater living beings.

The gauge for the new water cosmetics necessity was 1,000 liters of new water per liter of oil, with a scope of 200-2,000 liters of idealistic cynical cases. This outcome proposes that the requirement for sweet water make-up is critical when the saline water for algae development is utilized. In any case, the utilization of water substitution relies upon profitability cutoff points and saltiness of algae utilized as a part of the atmosphere, and different vulnerabilities and differences that have not yet been settled. Wastewater additionally is utilized as a part of algae development, accordingly decreasing groundwater and surface water utilization and waste water treatment by lessening nitrogen and phosphorus content. The potential favorable circumstances and constraints of wastewater usage to create biofuel algae are profitable. It has been reasoned that the development of double utilize microalgae for wastewater treatment and biofuel generation can possibly use wastewater supplements and decreasing the measure of new water expected to deliver natural biomass fuel.

Scale-up Considerations

It requires biofuel creation algae will be high and if algae biofuels are utilized to supplant a noteworthy part of the yearly US fuel transportation utilization, particularly in the event that you utilize untamed water tanks for algae development. In the event that you utilize open tanks for algae creation, a lot of water is expected to supplant the dissipation misfortunes of the tin surface and to forestall collection of broke down salt in the biomass culture framework. Late gauges from the US Department of Energy recommend that water misfortunes in the request of a few hundred liters of water for each liter of ocean growth or biodiesel oil would come about because of the utilization of tanks open in the dry and bright expresses the terrain of the United States. The most idealistic situation introduced in DOE generation (2010b) went toward the southwest of the United States. These evaluations depend on high rice generation rates (31 grams for every square meter [g/m2] every day) and high in normal cell oil (50 percent in dry weight). Given the climatic conditions, if algae are developed to create 220 billion liters of algae biofuel vanishing lake misfortunes would be 312 trillion liters for every year. This sum is about twofold the measure of water utilized for water system in the United States (177 billion liters in 2005, USGS, 2012). On the off chance that algae constraining farming territories with high precipitation, for example, zones near the Gulf Coast, the Great Lakes, and the greater part of the eastern drift, new water utilization per unit of fuel can be diminished by 75 percent. Little (2007) gauges that in the vicinity of 16 and 120 trillion gallons (60 and 454 billion liters) of water every year would be required for the creation of ocean growth oil expected to deliver 60 billion gallons (227 billion liters) Year biodiesel. The measure of fuel oil devoured by the US transport area in 2010 was around 207,000 million gallons (783,000 million liters) every year.

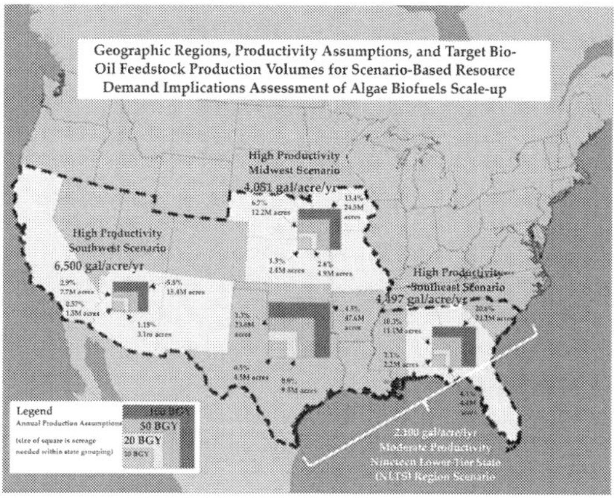

Figure 3.10: Scale-up scenarios for Algal biofuel forUnited States (four different regions).

SOURCE: Pate et al. (2011).

NOTE: 1 gallon = 3.78 liters; 1 section of land = 0.40 hectare; 1 gallon for each section of land = 9.5 liters for every hectare.

Little (2007) gauges that in the vicinity of 16 and 120 trillion gallons (60 and 454 billion liters) of water every year would be required for the generation of ocean growth oil expected to deliver 60 billion gallons (227 billion liters) Year biodiesel. The measure of fuel oil devoured by the US transport area in 2010 was around 207,000 million gallons (783,000 million liters) every year.

An examination assessed the aggregate sum of crisp water expected to reach, separately (37,800 million euro, 75,700 euro and 189,000 million liters), biodiesel target creation levels of 10,000 million euro, 20,000 million and 50,000 Millions of gallons for each time of freshwater algae for four areas of the United States. The assessments of the utilization of water for algae biofuels (Figure 3.10) fluctuate with the geographic district, the volume of crude material generation and efficiency target expected to achieve algae. Projections in the examination proposed that the utilization of freshwater algae creation frameworks by open lakes is probably going to be a noteworthy manageability issue if 10,000 million gallons (37,800 million liters) of biodiesel delivered every year, contingent upon locale. The development of freshwater algae 10,000 million gallons (37,800 million liters) of creation every year appeared to be unattainable in the Central and South-East districts in light of the fact that the water needs, which represent more than 70 percent and more than 170 percent of the water utilized for water system.

Despite the fact that it is evaluated that the water necessities for algal biofuels are higher than those of oil powers, practical utilization of new water ought to be considered with regards to territorial accessibility and other contending employments. For instance, an oil refinery situated in a parched district, groundwater revive is low and water deficiencies might be all the more harming to your nearby supply frameworks to develop algae in open tanks situated in a locale with An underground level, however the refinery devours considerably less water algae. The aggregate interest for nearby water assets for the creation of algae biofuels relies upon the administration practices of individual structures (for instance, the sort and measure of water utilized), the quantity of plants situated inside a bowl and the volume patterns and Time the volume of existing neighborhood aquifers, affected by aggressive water employments. The utilization of water and crisp and salt water withdrawals in the United States were evaluated by the United States Geological Survey.

It has been proposed by the specialist that water system water other rural applications ought to be occupied to the generation of algal biofuels on the off chance that they need to deliver 10 billion gallons of fuel from freshwater-developed algae. By occupying water system water from farming develop algae for fuel issues increment the utilization of sustenance water and fuel powers against. The abnormal state of reflection of surface water or groundwater associated with the surface of water frameworks can influence biological systems. Numerous biological systems require negligible occasional streams to help angle life cycles and riparian vegetation. Group of macroinvertebrates of conduits and assorted variety are likewise influenced by the present stream. In a few districts, the profundity of groundwater can influence the piece of earthly vegetation and cycle of supplements. The impacts of water withdrawals for algae development in natural populaces and biological community procedures would be vital to be considered in conjunction with water system, pressure driven, mechanical water utilize and metropolitan water utilize.

The generation of biofuels utilizing freshwater algae not sweet, for example, beach front ocean water are required; Waste water from agrarian, common and modern sources; Salty or salty bowls; and oil wells delivered from water, gas and coal-methane. We have to create viable ways to deal with diminish water misfortune by dissipation and to address saltiness development. These methodologies are more imperative for indoor situations where dissipation and saltiness gathering is required to be higher in the waterfront beach front sea with high relative moistness.

Sustainability Indicators

Supportability ramifications of water utilize are hard to measure. Many investigations utilize water utilization as a measure. Utilization is the expulsion and ensuing "misfortune" of surface water or surface by vanishing or surge. The connection between water utilization and the impacts of supportability, for example, environment change or deficiency of human needs, relies upon neighborhood conditions. As water supply is constantly underlined, there is an expanding requirement for strategies for connecting the different employments of water manageability impacts. Supportability pointers for new water for the creation of ocean growth biofuels include:

Consumption of new water communicated in kg of water per kilogram of fuel created (biodiesel or ethanol) or liters of water per liter of fuel delivered.

Inverted Water Efficiency (EROEI), super joule per liter (MJ/L).

These markers permit general examinations between destinations, crude materials and creation innovations yet don't give data with respect to neighborhood supply maintainability. For this particular venture reason and site particular extra data for the aggregate water utilization of a plant in connection to the momentum site offer will be required contrasted with future solicitations for all reasons, including biofuel creation. For instance, a structure that is assessed to require 1 percent of supply accessible in a range that is not anticipated that would encounter noteworthy populace development or the expansion in rural interest for water will be more feasible than a structure that requires The 50 percent offer accessible in a region with a quickly developing populace or horticultural request.

Pointers are likewise utilized as a part of expansion to water utilization. Water extraction alludes to the measure of new water or the expulsion of the buoy. Utilizing green, blue, and dim follows is picking up enthusiasm for some exploration groups. Green water is water dissipated amid creation, for example, developing harvests. Blue water is dissipated water system water amid edit development. Dark water is the measure of water expected to weaken contaminants by a procedure to meet the water quality principles. Picking which of these markers involves face off regarding; it is critical for scientists to report crude information on water utilization notwithstanding the intricate aftereffects of the pointers.

Nutrients

Algae require fundamental supplements for support and development of digestion, as on account of land plants. The exact stoichiometry of essential algal cells shifts starting with one condition then onto the next and between various types of algae. Be that as it may, photoautotrophic utilization of photosynthesis of algae to change over light vitality into new algal biomass with medium stoichiometry by and large because of the accompanying condition:

$$106CO_2 + 16NO_3^- + HPO_4^{2-} + 122H_2O + 18H^+ \leftrightarrow C_{106}H_{263}O_{110}N_{16}P + 138O_2$$

Rearranging Equation, the elemental content of algae can be expressed more simply as:

(CH_2O) $106(NH_3)16(H_3PO_4)$

The algae phosphorus nitrogen charcoal stechiometry can be considered as normal C106: N16: P1 moles, a valuecommonly known as the Redfield rate. As theorized in the condition, CO2 is basic for the photosynthetic generation of algal biomass, giving natural carbon that is required for cell combination of natural biomolecules, including lipids and starches may end up noticeably fluid biofuels. Nitrogen-containing atoms are engaged with catching and discharging vitality, cell structure and digestion. Phosphorous atoms are basic for vitality exchange. Both nitrogen and phosphorus are fundamental segments of ribonucleic corrosive (RNA) and desoxyribonucleic corrosive (DNA) hereditary polymers. It is fundamental to give CO2, nitrogen and phosphorus supplies that are incidentally adequately stable for the creation of extensive scale algae biofuel generation. In the event that developed species are ruled by taxa requiring silicon (Si), as diatomaceous, at that point it ought to give satisfactory if. Not at all like water needs, it can be normal that the measure of supplements required advancing algae development is the same if open or shut bioreactors are utilized. Notwithstanding, supplements can be reused all the more effectively in shut photobioreactors in open tanks. The degree and productivity of the reusing of supplements utilized as a part of the consequent preparing of algal biomass in biofuels and co-items influences the net necessities of nitrogen and phosphorus biofuel creation frameworks. Supplements cannot be reused to 100 percent because of precipitation misfortunes and supplements identified with algae biomass passing. The development of these suspended mucks and broke down natural substances that goes with a sink for supplements and decreases the accessibility of light for the development of algae that live. The dead biomass of algae cannot be left in the lake to mineralize because of undesirable results for the way of life medium.

Understanding CO2 limitations and necessities is basic to tending to the profitability of crude material development (ie biomass or lipid yield contingent upon the preparing way used to deliver fuel). Profitability in crude materials, thus, influences the financial feasibility of algal powers. On the off chance that generally immaculate additional ocean growth streams require CO_2, recognizing the likelihood of developing algae close to this asset could lessen transport costs. Concentrates distributed on algal biofuels recognize that the generation of effective algae requires outside wellsprings of CO_2 think (provided in vaporous shape or bicarbonate), likewise in light of the fact that the energizing rate of CO_2 frameworks from algal culture can be constrained by dissemination through the interface air-water.

From the arranging perspective, ideal biomass must be accomplished by air circulation or supplementation. As the air CO_2 fixation is expanded, amphibian researchers have started to evaluate the potential impacts of CO_2 focus on algae efficiency and stoichiometry. In the greater part of the algae, the photosynthetic movement of the essential ribulose-1 compound, 5-bisphosphate carboxylase oxidase (Rubisco) is lower than the soaked CO_2 level in adjust with the environment, recommending that CO_2 is a constraining variable of carbon settling photosynthesis.

For instance, utilizing test controls of CO_2 focus in covering boreal lakes, The essential phytoplankton generation rates were up to ten times higher in water covering lake with CO_2 than water containing CO_2 at harmony fixations. Moreover, the coordination of algae generation frameworks with CO_2 can possibly build the yield of algal biomass per unit of nitrogen and phosphorus in algal cells, particularly in low light conditions, along these lines expanding the productivity of utilization of nitrogen and phosphorus. Indeed, even a direct increment in CO_2 could conceivably expand C: N and C: P in the algae. In a current substantial scale, three-digit CO_2 try past current climatic fixations has been converted into C: N proportion of algae (8.0 moles) which has surpassed Redfield stoichiometry.

Estimated Nutrient Requirements:

- Carbon Dioxide:

Assessed CO_2 prerequisites for the generation of algal biofuels are generous. For instance, Pienkos (2007) gauges that 14 to 21 kg of CO_2 must deliver algal biomass expected to make 1 gallon of biodiesel (3.69-5.54 kg of CO_2 for every liter). A specialist evaluated that 14-35 kg of CO_2 is expected to deliver 1 gallon of ocean growth oil (3.69-9.23 kg of CO_2 for each liter of kelp oil). The two esteems can be contrasted with gauges made by algae biofuel show offices. For instance, Algenol gauges that 734 tons of CO2 would be expected to create 100,000 gallons (378,000 liters) of ethanol every year in its photobioreactors (DOE, 2010a); this equivalents 1.94 kg of CO2 for every liter of ethanol or 2.91 kg of CO2 for each liter of proportional fuel. Sapphire Energy gauges that in the vicinity of 15,000 and 30,000 tons of CO2 every year would have been utilized as an added substance to advance 30,000 barrels (4, 78 million liters) of ozone harming substance algae development (USDA-RD, 2009); This compares to 3.14 to 6.28 kg of CO2 for each liter of raw petroleum or equal gas, accepting 1 liter of unrefined petroleum can be updated for 1 liter of oil fuel equal.

- Nitrogen and Phosphorus:

It has been examined the prerequisites of nitrogen and phosphorus from a generation framework utilizing open tanks for the development of algae to deliver algae oil. They took Redfield C: N: P stechiometry in the creation of algal biomass. For straightforwardness, it assimilates 100 percent supplement ingestion proficiency and does not consider the possibly huge commitments of nitrogen and phosphorus expected to balance the wasteful aspects and misfortunes of a biofuel generation framework. Reusing of supplements is not taken in the projection of its models. They ascertain these suspicions, anticipating that each ton of dried biomass algae created by the required framework is 88 kg N and 12 kg P. Expecting half oil containing algal biomass, these nitrogen and phosphorus prerequisites are 0.61 kg and 0.083 kg N Per gallon of oil P created algae (or 0, 16 N and 0,022 kg P for every liter of kilograms of algae oil delivered). Nonetheless, if 20 percent of ocean growth oil is assumed, these dietary prerequisites increment to 1.5 kg and 0.21 kg of nitrogen phosphorus per gallon of oil delivered (or 0.40 kg N and 0.055 P kg for every liter of Algae oil created). Comparable evaluations of 1.1 kg and 0.24 kg of nitrogen phosphorus per gallon of biodiesel delivered (0.29 kg N and 0.063 kg P for every liter of biodiesel).

It has been recommended that immediate ethanol blend may have less nitrogen and phosphorus necessities than the biomass cyanobacter reference receptacle is not expelled from the generation framework amid biofuel collecting. Along these lines, just nitrogen and phosphorus for the development and upkeep of essential biomass feet; Nutrients joined in algal biomass are not lost amid catch of ethanol. It has been evaluated that 56,000 liters of ethanol for each hectare every year compare to a nitrogen and phosphorus necessity of 0.065 g N/MJ and 0.0024 g P/MJ (0.002 kg N and P 0.0001 kg for each liter of equal petroleum).

Scale-up Considerations

- Carbon Dioxide:

Human-centered CO2 delivered can be utilized as a part of the generation of algae biofuels. For instance, coal-let go smoke is a potential wellspring of CO2. The potential advantages of having algae generation frameworks, stationary CO2 sources and potential hindrances for modern utilize are examined in the National Road Technology Biofuel Algal Research Report. This report is likewise incorporated into a guide of wellsprings of CO2 from the plants situated inside 32 kilometers of metropolitan wastewater frameworks of the favored climatic districts distinguished in the assessment of the expansion of the Sandia research centers. On the off chance that CO2 settled CO2 situating is not practical then CO2 catch and transport expenses ought to be considered.

Under the situation theory, about portion of the emanations of settled sources in 19 bring down level states would be expected to give adequate CO2 amid sunshine hours to help the generation of 10,000 million gallons (37,800 million liters) every time of ocean growth oil. For all other stage districts 90 to 150 percent of all wellsprings of CO2 outflows would be expected to deliver 10 billion gallons of algal oil. They presumed that exclusive few settled source discharges would be in a sensibly helpful get to extend in US locales that are most appropriate for expansive scale algal biomass development, unless there is a costly framework for catching and CO2 pipeline. Pienkos (2007) evaluated that it would take 0.900 million to 1.500 million tons of CO2 for the creation of kelp oil for the significantly bigger focus of 60,000 million gallons (227,000 million liters) of biodiesel every year. This CO2 request rises to 36 to 56 percent of current CO2 outflows from all US plants.

On the off chance that CO2 supplementation is required for elite algae generation, the measure would be restricted to algal biomass; it can be monetarily created in the United States. Besides, as verified by a specialist 2011, future duties regarding decrease barometrical CO2 emanations under the Kyoto Protocol recommends that there could be a generous move towards the era of low-CO2 power, for instance, utilizing sustainable power source or power produced by Charcoal with carbon catch and carbon stockpiling. De-carbonization of power creation diminishes the quantity of areas where fumes gas can be monetarily accessible for the development of algae for biofuels. Different sources and inorganic carbon frames, for example, the supply of strong bicarbonate, could be created for the generation of advantageous biomass scale autotrophic microalgae mode.

- Nitrogen and Phosphorus

It has been inferred that it is important to help four levels of ocean growth biodiesel delivered from algae developed in four diverse US geographic areas anticipated for nitrogen necessities for the creation of 10 billion gallons for every year (37.8 billion liters

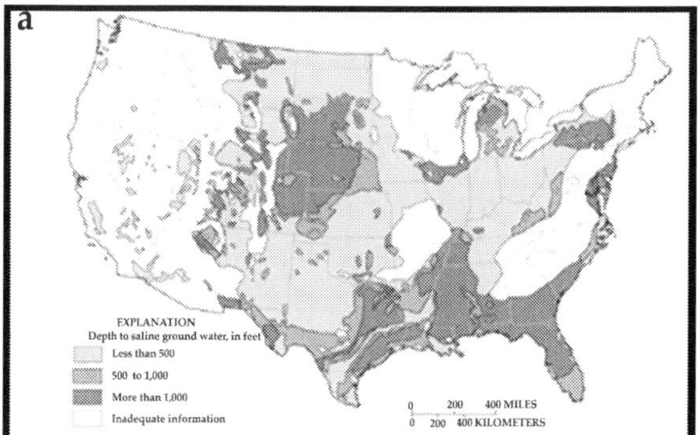

(a) Depth tp saline groundwater (units)

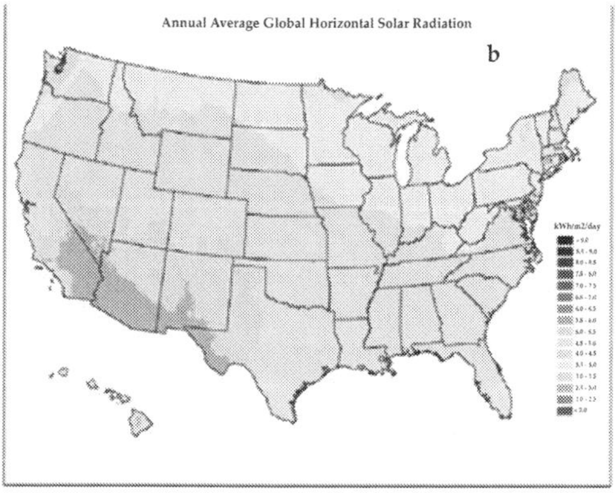

(b) Annual average solar radiation (units);

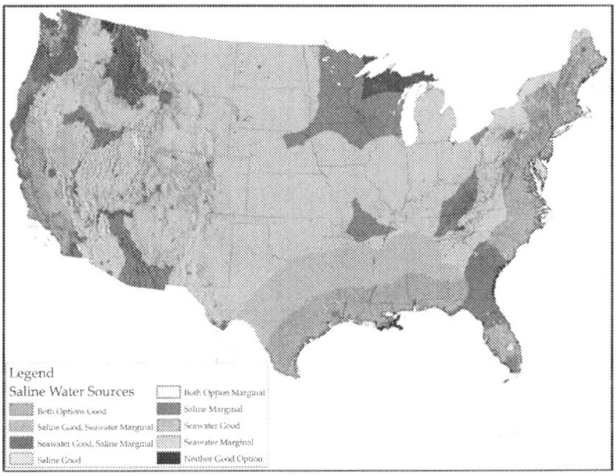

Figure 3.11: Saline assets in the mainland US, in view of the generation of evaluated potential vitality versus the vitality required to transport the water.

SOURCE: https://pubs.usgs.gov/fs/fs075-03/

Finish point by point life-cycle appraisals (LCA) and a natural effect evaluation before any huge scale organization are valuable to and vital for guaranteeing a smooth way to commercialization, especially if the land being considered for siting has not ever been produced. Moreover, the destinations where algal development frameworks can be introduced will be compelled by high land cost, horticultural action, ecological esteem, and inborn social estimation of the land being considered. Biofuel-driven land-utilize change (LUC) additionally conceivably could make critical natural effect and supportability concerns.

A preparatory abnormal state appraisal of microalgal biomass creation potential was performed by Sandia National Laboratories. The atmosphere criteria utilized as a part of this spatially unequivocal investigation were: yearly normal aggregate sun hours of 2,800 hours or above, yearly normal every day temperature of 55°F (12.8°C) or higher and yearly normal stop free days of 200 days or more. As far as satisfactory daylight and appropriate atmosphere, parts of Hawaii, California, Arizona, New Mexico, Texas, Louisiana, Georgia, and Florida seemed, by all accounts, to be the most encouraging in the United States. Northern states, for example, Minnesota, Wisconsin, Michigan, and the New England states experienced solid occasional inconstancy in protection and temperature. The evident absence of engaging quality of the Gulf Coast district in the Sandia think about was ascribed to bring down yearly normal sunlight based protection. Conversely, another examination by the Pacific Northwest National Laboratory proposed that yearly productivities in this district were higher than those evaluated by the Sandia think about. Nonetheless, the forecasts by Sandia National Laboratories are reliable with the guide of potential algal oil creation for open-lake generation frameworks.

At the site level, nearby geography and soils additionally can conceivably constrain the accessibility and reasonableness of land for open-lake development. The establishment of huge shallow lakes requires level landscape that has an incline of close to 5 percent on account of the characteristic needs of the innovation and in view of expanded expenses of site advancement and liquid pumping. Neighborhood wind conditions influence blending and temperature in lakes and the honesty of lake liners. High breeze in a dusty or sandy area could add to silt stacking to open-lake development frameworks, which may require visit, tidy up.

Siting biofuel generation frameworks near water and supplement assets would put extra breaking points on algal biofuels commitments to future fluid transportation markets. They noticed that even where arrive appropriate for extensive scale algal biofuel creation exists, the nearby accessibility of basic assets could influence the financial matters of generation, net vitality return, and GHG discharges. The ideal locales for business scale algal biofuel creation would have either the required assets in nearness or components set up to guarantee sufficient and continuous supplies of these assets. Specifically, access to substantial volumes of new water, saline water, or both will be basic for algae development (Figure a,b). An analyst communicated worry that the accessibility of adequate supplies of supplemental CO2 is questionable (Figure c) in the land areas that are most appropriate to year-round algae development.

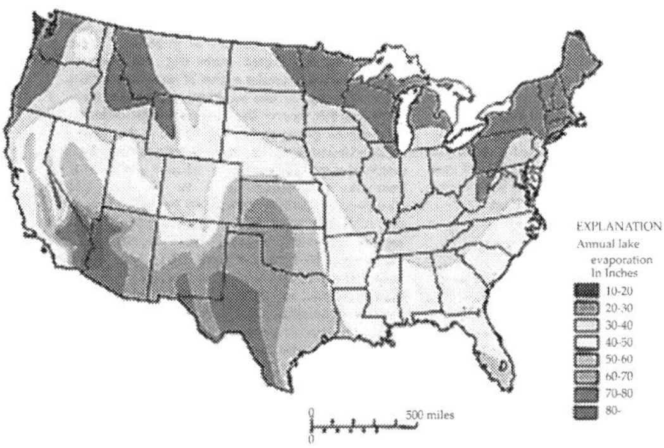

Figure 3.12: Per annum average lake evaporation in the continental US

SOURCE:http://www.islandnet.com/~see/weather/almanac/arc2010/alm10jul.htm

Notwithstanding the photosynthetic surface region vital for algae development, assessments of the aggregate land required for aalgae development office would incorporate the space required for inoculums development; frameworks for conveying inoculums to development vessels; reaping frameworks; repositories for holding water; squander administration, stockpiling, and reusing offices; and other emotionally supportive networks. For instance, bowls might be expected to hold dilute discharges from blow in an open-lake framework. Also, extra land for berm development is required if the raceway lakes are built with earthworks. Murphy and Allen (2011) represented the land required for framework to help the essential development offices by utilizing a scaling component of 1.6:1 while evaluating the aggregate range load for open raceway lakes water assets (for instance, aquaculture lakes and waterfront zone waters) likewise could be considered. What's more, algal biofuel offices could be sited on lands that are profoundly appropriate for sun based vitality creation, for example, bone-dry grounds in the southwest or radiant beach front areas. Be that as it may, deficient data is right now accessible to survey the suitability or creation limit of these potential pathways. Additionally, potential tradeoffs (for instance, water accessibility) would need to be considered.

Supplements of ocean growth wastewater evacuation can be consolidated with the creation of biofuels. For instance, more than 15,000 local waste water treatment plants in the United States all in all deliver around 34,000 million gallons (128,700 million liters) of profluent squander water every day. A few thousand some little (under 10 hectare) and some vast scale (more than 100 hectare) framework right now working for the treatment of urban wastewater.Food preparing offices and dairy homesteads and control are different wellsprings of fluid waste rich in supplements. Operations, for example, these possibly give alluring arrangement locales where wastewater could be utilized as a wellspring of water and supplements for incorporated biomass algae generation frameworks. One issue is that it once in a while has enough arrive on EDAR destinations in urban ranges.

Two sorts of frameworks can be seen: Dedicated algaculture conceivable outcomes, whose primary reason would be the generation of algal biomass (such frameworks additionally require wastewater treatment and supplement reusing); And decontamination plants, the principle motivation behind which would be the treatment of wastewater, yet the creation of algal biomass which can be gathered because of the treatment procedure. It is normal that such incorporated frameworks will have many favorable circumstances, including giving a wellspring of minimal effort supplements, decreasing interest for other new water sources, potential cost of diminishing wastewater, enhancing the nature of treated effluents and lessening potential Greenhouse gas discharges. Be that as it may, it is not clear what number of locales of sufficiently potential surface bordering adjoining soil would be appropriate for extensive scale algal biomass development.

Other creative site endeavors can be legitimate if there are satisfactory supplies of light and different assets. For instance, aquaculture is a vast division in the United States; numerous catfish creation lakes are out of generation and along these lines are conceivably accessible for elective use for the creation of algae biofuels. More than 100,000 sections of land of catfish creation lakes may just be accessible in the territory of Mississippi if these current structures would be reasonable for algal biomass generation and collecting.

Estimated Land Requirements

The generation of algal biomass possibly requires considerably less surface than the development of crude materials for soil biofuels. Earth necessities for the creation of algal biofuels are possibly 1-2 orders bring down in estimate than any harvest biofuel construct either in light of the premise of volumetric effectiveness or vitality proficiency per unit of surface. Future upgrades in algae oil substance and efficiency of algae cultivating frameworks (kg/ha every year) are required to additionally diminish the measure of soil expected to deliver a given measure of fluid biofuels. A national appraisal of the conceivable generation of algae biofuels and their utilization of assets in open tank frameworks was directed by an analyst. The appraisal depended on a hypothetical plant that comprises of 100 profundities of 30 feet and 4 sections of land that requires around 400 hectares of land for similar lakes and 90 hectares to oblige extra operational foundation. The capability of ocean growth oil and the related land and water sources prerequisites, in light of the prevailing physical procedures that impact the development of algae creation which have been reenacted. Water, supplement and CO2 supplies were not constrained to reproductions to give hypothetical appraisals of yearly normal yearly microalgae vast water tanks and, if all locales have created fitting ground and water prerequisites.

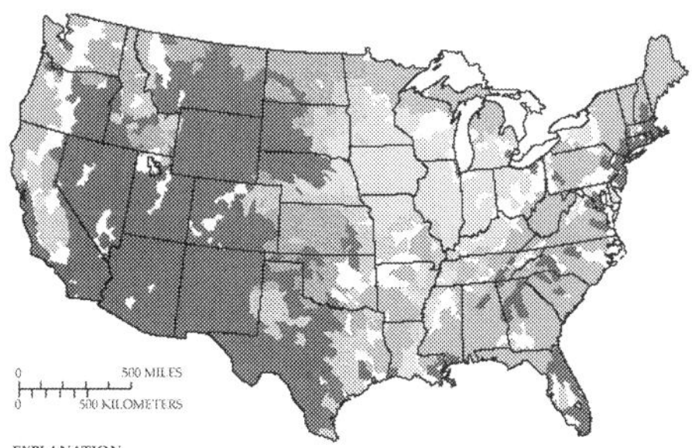

EXPLANATION

LAND USE SHOWN ON MAP	NITRATE 1980-89		TOTAL PHOSPHORUS 1982-89		SUSPENDED SEDIMENT 1980-89	
	Yield, in tons per square mile per year	Percentage change per year	Yield, in tons per square mile per year	Percentage change per year	Yield, in tons per square mile per year	Percentage change per year
AGRICULTURE Wheat	0.032	*	0.010	-2.8	10	+0.8
Corn and soybeans	0.932	*	0.163	-2.1	100	-1.0
Mixed	0.304	*	0.066	-1.6	79	-0.7
URBAN	0.547	+0.2	0.119	-0.6	23	-0.6
FOREST	0.255	*	0.063	-0.8	31	-0.3
RAN GE	0.031	*	0.017	-1.9	33	-0.2

Figure 3.13: GIS based analysis showing approximate per annum production of the algal oil (L/ha per year).SOURCE:https://water.usgs.gov/nwsum/sal/trends2.html

An examination distinguished 11,588 aggregate non-aggressive 430,830 km2 fields that could be utilized for the generation of extensive scale algae and open tanks. Solid geographic variety has been anticipated into its yearly biofuel creation levels (L/ha every year) (Figure 3.13), the accessibility of soils and new water supports the spots in the Gulf Coast district. The most noteworthy yearly normal generation rates happened in focal Texas and quite a bit of Florida, however arrive costs in these locales could influence the reasonability of its utilization for algae development. Their examination recommended that in view of the speculations of its present investigation and innovation, algae can possibly be developed on a huge scale to 5.5 percent of the land in the commensurate United States to produce 220 billion liters of oil for each year. This measure of ocean growth oil is proportional to 28 percent of aggregate oil utilization in the United States for transport. The same resaercher likewise led quantitative appraisals of regional cases emerging from the era of biofuels from algae and inferred that dirt prerequisites were the most reasonable among the asset (water, supplement and soil) necessities considered in their examination. As on account of water needs, arrive requests fluctuate with the geographic area, the measure of crude material to be delivered and the efficiency level of the algae is thought to be accomplished. They reasoned that the locales of the southwest and nineteen lower level states in the United States were conceivably more inclined to meet the prerequisites of executing the creation of algae biofuels on a bigger scale in the Midwest and Southeast areas. In any case, it is normal that the aggregate surface range expected to reach biofuel generation levels is conversely connected with yearly biomass profitability and algae lipids that can really be acknowledged practically speaking (see Figure 2). A specialist utilized an itemized building model for Nannochloropsis photovoltaics on a mechanical scale developed for the count of a necessity of 4.41 million hectares for the generation of 40,000 million liters of biodiesel got from microalga every year. That land range identical to 16 percent of the aggregate land territory in Colorado and 0.45 percent of the US surface.

Sustainability Indicators

Maintainability markers soil necessities for the generation of algae biofuels include:

- Liters of fuel created per hectare.
- Kilograms of fuel created per hectare.
- Energy return ashore contributed (MJ/ha), comparable to EROWI.
- Land required meeting different generation targets in respect to existing national use or supply.

The meaning of markers ought to determine whether the vitality expected to get ready land and to transport assets to the site is incorporated. Ground markers depend, all things considered, on the area of the earth, albeit every one of the pointers have in any event a few specifics of the site. The DOE could help fathom these markers through more nitty gritty investigations of potential biofuel generation limit of particular advancements in various districts of the nation.

To abuse the high photosynthetic proficiency of algae, vitality should be put resources into horticultural frameworks and bioraffines to develop algae, oversee applications and process water algaes in the coveted fuel. Given the utilization of critical vitality in the energizes of different biofuels, examination of the whole fuel cycle of algae funnels is fundamental to understanding the vitality suggestions. This segment looks at the condition of learning of the vitality properties of biofuel generation frameworks.

CHAPTER 4

ENVIRONMENTAL IMPACT OF ALGAL BIOFUEL

ENVIRONMENTAL EFFECT

Production and use of biofuels have benefits and side effects. This chapter analyzes the potential for production and use of algae biofuels, environmental impact, potential or real acceptance of the perception of the effects of the company's influence and some of the impacts on health resulting from the specific environmental effects. Water quality, greenhouse gas emissions (GHG), air quality, biodiversity, waste effects in genetically modified algae.

As far as possible, this chapter of algae biofuel life is analyzed, the potential of algal biofuels to improve sustainability issues with respect to fuel and other biofuels and the potential to mitigate adverse effects throughout the cycle. Environmental sustainability indicators and data collected to assess sustainability are suggested. In some environments that respect environmental management systems and biofuels to assess environmental performance, they are easy to measure and are basic databases, but this does not occur in all systems.

A number of potential environmental concerns are obvious, and if concerns are not addressed, it could become a significant risk in large-scale distribution. As in any other industrial or agricultural enterprise, once recognized, such risks can be managed in accordance with international standards or industry is forced to reduce the effects to acceptable levels. For the sake of completeness, several potential environmental hazards in this chapter are mentioned, but some are less likely than others. Some of the environmental risks may require an exploratory evaluation and subsequent monitoring to ensure that they do not become sustainability issues if they increase the production of seaweed biofuels

WATER QUALITY

Water is likely an important limiting factor for the growth of algae. To expand the growth of cultivated soil algae requires; fortunately, many of these regions have alkaline water deposits or substantial saline below them, providing a significant source of non-drinking water that is suitable for the growth of many species of algae. Perhaps surprisingly, algae grown in open ponds have water requirements similar to cotton or wheat area, but less maize, to replenish evaporated water. It is essential, given the large distribution of algae; it is advisable to use water to avoid a debate, "water against fuel". Although there are substantial alkaline reserves, water will remain central to the production of seaweed biofuels and should be carefully considered as the industry expands.

Culture and Process Water

It is likely that water for algae cultivation is recovered and reused to reduce the consumption of water and water. Liquid effluents can also be recycled to anaerobic digestion of lipid-extracted algae to produce biogas. Whether to release water collection rather than anaerobic nitrogen recycling (N) and phosphorus nitrogen (P), the concentrations depend on nitrogen and phosphorus absorbed by the collected algal biomass. The released water may be more saline than the receiving water, especially if the algae salty culture used is used. This discharge point is governed by the Clean Water Act and should obtain the elimination of permitted national emissions of pollutants to operate the ability to grow algae. However, the violation of the permit has been observed in some refineries using crops of terrestrial biofuels as raw material. Regulatory Compliance and Assurance would address the concerns about water collection.

Eutrophication

Potential Environmental Effects

Growing large-scale algae requires the supply of large amounts of nutrients, especially nitrogen and phosphorus, to ensure high performance. Even when nitrogen and phosphorus are not excessive, total nutrient concentrations in algae biomass will be high. Despite the unlikely accidental release of culture water in surface water and soil, this event could eutrophicate sea and freshwater valleys, depending on the proximity of algae ponds, soil surface, and precipitation.

High concentrations of inorganic nutrients, particularly nitrogen and phosphorus in a water reserviour stimulates the growth of algae, hence, results in the excessive algal biomass production. This phenomenon is called eutrophication. As algae die and decompose, high levels of matter and organic decomposition processes reduce water oxygen and lead to anoxic conditions. In some cases, changes induced by eutrophication may be difficult or impossible to reverse if they can produce alternative stable states in the affected ecosystem. The effects of eutrophication have been well studied and depend on a load of nutrients in water and water time to receive and the volume of water in these systems. The nutrient feed may lead to anoxia in the cold portion of cold lakes or hypoxia in water bodies. Potential biotic eutrophic effects include changes in the density and structure of algae and biomass of the largest ecological systems. Fish yield is influenced by phytoplankton biomass and phytoplankton salinity levels.

The level of nutrients in estuaries and coastal marine waters play a signifinact role in determining the productivity and the structure of the community of primary producers and, by extension, the productivity and structure of higher tropical levels. Nutrient-rich systems tend to have a small marine herbal community. Eutrophication also has implications for the social acceptability because of the aesthetic concers that affect the recreational value of water bodies. If water releases a rare culture or the physical appearance of algae growth could have negative effects on the social acceptability of unbiased stagnant biofuels.

Opportunities for Mitigation

Quantifying water loss from canals, ponds or photoreactors indicates that repairs are necessary if small losses. These culture systems can be designed and tested to withstand natural disasters are possible throughout the life of the infrastructure. In coastal areas, for example, designs and infrastructures should take into account the likelihood that winds and waves reach the site for algae production. It would be desirable mitigation plans for accidental releases. External agriculture Thealgae can also be located in places that are not prone to hurricanes or far from lakes and canals. With regard to water collection, engineering solutions can maximize recycling.

Waterborne Toxicants

Possible Environmental Effects

Herbicides are utilized to control the development of macrophytic algae in the algae lakes or photobioreactors. The expansion of herbicides may make harmfulness the uncovered people and different life forms. There is likewise a plausibility of overwhelming metal tainting when the water utilized for the development of algae is released into the encompassing condition. It is additionally realize that numerous types of algae including cyanobacteria, diatomas and Chlorophyta have the capacity to bioaccumulate the overwhelming metals. Contingent on the kind of herbicides and source water utilized the contaminants may constitute large amounts of natural mixes, oils, and fats, boron and alkali (NH3), mercury and carbon dioxide and so forth. In this way, it is of key significance to distinguish the potential dangers related with re-utilizing of water for water system purposes, utilized amid development and gathering of algae fills.

Opportunities for Mitigation

Checking of metals and different mixes from water sources, supplement sources, and culture showing plants give data on the way that harmful waters represent a huge issue. All things considered, specialized arrangements would be expected to evacuate poisonous conduits to maintain a strategic distance from proficient and natural exposures. Mercury is expelled from fumes gas in a few arrangements of carbon producing units. Mercury expulsion is ineffectual for specific sorts of coal and vegetable designs. Poisons in exhaust could put different confinements on the kind of coal-terminated power plants would be appropriate to give CO2 algae.

Groundwater Pollution

Possible Environmental Effects

Open tanks may not be reasonable for some sorts of territory without the utilization of a covering, and the requirement for a full survey of potential surface and groundwater impacts if covered earth lakes are utilized. In the event that open air lakes have misaligned imperfections or wear safe coatings, water holes may happen in the nearby fumes framework. Dirt is delegated reduced basic trustworthiness and can be contrasted and manufactured coatings. Be that as it may, trustworthiness can be traded off by poor development. Nitrate draining has been seen in organized earth soils, yet the subjective appropriateness of dirt outcomes is obscure. Nearby earthly vegetation could assimilate a portion of the way of life media discharged by invasion. In a few territories, if open lakes contain high convergences of broke down inorganic nitrates, filtration can add to concerns identified with nitrate harming if groundwater is utilized for drinking animals or for people.

Evacuate Salt or salt water may prompt salt water and salt water if the crisp ground is not all around isolated. Salinization of groundwater is a capability of some farmland where water system is a typical issue. Nonetheless, one of the fundamental favorable circumstances of ocean growth biofuels is that crude material can prompt non-cultivable soils, so it is impossible that salinization of farmland to gather new water for algae development.

Opportunities for Mitigation

The utilization of fixed culture frameworks for all intents and purposes disposes of the potential loss of algae, except for disastrous crevices. Utilizing open frameworks could actualize innovations, (for example, the improvement of long haul waterproofing and covering frameworks) and local answers for limit supplement misfortune, and could create principles to limit misfortunes. For instance, Phyco Biosciences utilizes a station with a lightweight covering. It is required to take out or limit spillage misfortunes beneath 0.01 percent. Conceivable preventive measures could incorporate soil particulars, joined with a profundity characterized by the base of the tank to the groundwater esteems. Additionally, nearby directions are probably going to require secured lakes, which would decrease the probability of water spills, yet add to capital expenses and prompt impermanent bans when coatings have been supplanted because of wear or disappointment. Measures to forestall incidental water release (eg passageways or flood bowls) amid extraordinary climate occasions could help avoid water contamination.

Wastewater Treatment

The primary errand of wastewater treatment is the expulsion of nitrogen and phosphorus. In conventional treatment frameworks, phosphorus is especially hard to evacuate. In the treatment of synthetically hastened propelled wastewater utilizing artificially based aluminum-or-ferrous coagulant phosphor to shape an insoluble strong, or water is evacuated by microbial action. The phosphorus is then recuperated covered in a dump or slime treated to make compost.

Reused wastewater nourishment and its utilization for the creation of supplemental algae could be an intriguing choice for the utilization of other disposed of supplements. It has been discovered that algae medications are as productive as a compound treatment in expelling phosphorus from the wastewater. Additionally, as algae biomass gathers supplements that have been ingested amid cell development, coordinated septic frameworks can give an imperative nourishing cure benefit. In research center scale tests, more than 90 percent nitrogen and 80 percent phosphorus reflux have been generally disposed of by the algae Chlorella vulgaris. Correspondingly, Chlorella and Scenedesmus research center societies expelled from 80 to 100 percent NH3,

Algae can possibly expel supplements from rural or mechanical wastewater. A few examinations have discovered high efficiencies in expelling nitrogen and phosphorus from squander water containing wastewater and this could likewise be utilized as contributions for Biofuel Plants. Algae biofuel frameworks can possibly enhance water quality and advance city or horticultural wastewater treatment frameworks with more noteworthy manageability. In any case, keeping up lipid-rich strains and controlling development conditions to advance lipid generation have not been reliably shown to outside lakes, lakes including wastewater treatment. Modern wastewater has bring down supplement focuses and higher harmful fixations, so they are less defenseless to being utilized to create algal biomass required for business biofuel generation.

Incorporated algae biofuel creation frameworks can evacuate numerous different contaminants, for example, metals and natural contaminants, including endocrine disruptors. In the event that the retention of algae contaminations is alluring relies upon whether coproducts are to be delivered with energizes or lipid removed biofuels to be utilized for supplement reusing. Expulsion of contaminants from these frameworks enhances the nature of water however can likewise represent a potential risk if transitory oceanic life forms utilized straightforwardly or inadvertently high-algae in human culture as creatures have been constantly presented to drying of algal biomass for assimilation Contamination of poisons is undesirable algal biomass if remaining or creature corrective item is utilized.

Comparison of Pathways

The models depicted in Chapter 2 influence the sorts, probabilities, and greatness of the impacts of water quality. For instance, moderate arrival of supplements in indigenous habitats (and more noteworthy potential for eutrophication and groundwater contamination) are regular for open frameworks, however not for shut frameworks. Herbicides are probably going to be utilized just in open frameworks.

Sustainability Indicators

Manageability markers for water quality incorporate watery fixations and heaps of supplements, herbicides, metals and salt water saltiness. These markers are standard measures for the nature of water and wastewater. Supplement fixations in connection to an execution or any unfriendly consequences for water quality (eg eutrophication) are incorporated. These are for the most part measured amounts, and can likewise gauge basal and normal inconstancy. Burdens are field estimations and recreation comes about speak to the commitment of the way of life of biofuel media tossed into beneficiary waters. These can be contrasted and different fillers for those waters.

- The grouping of nitrates in groundwater and conduits
- Total nitrogen fixation in streams, lakes, bowls, and estuaries
- Total phosphorus fixation in streams, lakes, bowls, and estuaries
- Nitrate stacks in waterways and aquifers
- Total phosphorus stream
- The groupings of herbicide streams
- Water based herbicides
- Concentrations of metals in waterways
- The groupings of metals in crops

LAND CHANGE

Potential Environmental Effects

Change of land is a change of human activity on the continent. Changes in land play a role in the sustainable development of algal biofuels due to associated environmental impacts such as greenhouse gas emissions, changes in biodiversity, and changes in ecosystem services such as food production. Moreover, there is growing social concern over the types of spatial and temporal conversion, such as deforestation and urbanization stairs. The impact on the development of algal biofuels will depend in part on the type of soil conversion, the land use area has changed, the intensity of land interruption and management, and the duration of the change (eg if reversible)

The commercial scale production of algae biofuels will require a considerable area for each installation, and the distribution of large-scale algae biofuels will lead to the conversion of land to other existing uses. Conversion gaps, processing plants and refineries for most products are found and the conversion of energy areas are connected, such as roads and power lines to the plant infrastructure, will be more generalized and involve linear structure structures. This section focuses on land use change (LUC) associated with algae culture, as the variation associated with plants or raw material refining does not differ from that of other liquid fuels.

Direct land change is generally defined as the cause and effect between biofuel development and land conversion in the absence of strong external mediation factors. Direct land change occurs within the course of biofuel production when the land is dedicated to being used for the production of biofuels. However, in practice, it is presumed that the direct transfer of land use for biofuel production includes land used for the production, processing, and storage and refining of raw materials. The indirect change in land use occurs when biofuel production results in new land changes in other parts of the country or in another country through market-mediated effects

Direct and indirect changes in the land could affect net greenhouse gas emissions of biofuels. Direct land use change can cause carbon capture or greenhouse gas emissions, depending on the conversion rate and the ground. For example conversion of annual plant production perennial crop production that can improve carbon storage in the chassis. In contrast, the plant grows native ecosystem cleaning to produce only one result to free a sack in the atmosphere of Jesus

Pastures and pastures could become algae farming, and the transfer of these algae uses may or may not have other indirect effects. If the lawn or grass is excessive and unused, then land reclamation is not subject to indirect land displacement (ILUC). On the contrary, if the planted algae propelled livestock pastures are made, the producers decide to change for the production of corn feed. Changing wheatgrass production of cattle fed foods also administered put pressure on the grain market. Alternatively, if herbaceous plants and meat production limit, to reduce the production of this dirt removal, then animal production fed with grass can be replaced elsewhere. Indirect changes in soil use do not only affect ecosystem services, but must take into account changes in greenhouse gas emissions in the assessment of greenhouse gas emissions from algae biofuels.

If the indirect effects of biofuel production from algae, then quantifying potential impacts of biodiversity, water quality and water balance include those associated with indirect ground conversions. The preventive quantification of the indirect effects of biofuels has been limited in general to the effects of greenhouse gases and the impact of food security.

As in the case of biofuel crops, it is difficult to determine indirect changes in land use from market estimates and associated greenhouse gas emissions are very uncertain. Although complex models have been used to project the extent of indirect changes in land use after bio fuel cultivated in the soil, the committee has similar projections for algal biofuels. Algae farming are less likely to incur indirect changes in land use because it requires high-quality agricultural land. Conversion of farmland into new crops (biofuels) will also require a new property or the will of farmers to cultivate a new product. Algae biofuel increase requires several work programs stacked in a row. Although agricultural land does not become algae, the discussion above of the potential conversion of pastures illustrates the potential for indirect land-use change.

Comparison of Pathways

As to the change in soil use, the main difference between the ways in which Chapter 2 is the difference between the reasons for open systems and water photoreactors. Spatial and temporal changes in land use would be proportional to the allocation of land.

Potential Opportunities for Mitigation

In general, the development of algae biofuels avoids forests and the value of agricultural land. Avoiding meadows and areas with high biodiversity or recreational value would also remove some of the sustainability concerns associated with the commercial development of algal biofuels.

Sustainability Indicators

Changing land has not been consistently proposed as a criterion for sustainability, though it is often considered a factor in sustainability aspects of biofuels (eg greenhouse gas emissions, biodiversity, water quality and soil quality). Consequently, some collections of sustainability indicators do not include indicators of sustainable land use. However, there are aspects of land use and infrastructure, waterproof surfaces, and some disturbances, which may be of long duration or irreversible and cannot be addressed adequately by other categories of sustainability indicators. Possible indicators include sustainable land surface ground impervious disturbance. Changes in the waterproof surface affect the water cycle and the dynamics of the basin as well as terrestrial and aquatic habitats. The disturbed territorial area can be considered as a measure of sustainability. Floor areas can be standardized on the basis of a ground factor condition (Equation 4.1) that captures the extent to which the aspects of development transformation, infrastructure, potential accidents, and basic use of energy variation of the state Natural and the ability to provide ecosystem services.

$$\frac{\text{Land Disturbance Area} \times \text{land condition factor}}{\text{Energy Output } (\text{or Volume}) \text{ of Liquid Fuel}} \quad (1)$$

It is vital to evaluate inclines in arrive utilize change identified with the creation of algae biofuels. Nonetheless, while there is no history of the business advancement of biofuel algae creation offices, extend alterations and likely land utilize patterns in view of monetary and social factors and burdening natural variables will be required.

Whenever critical or uncommon biological system administrations are given through the gauge, a measure of such administrations could fill in as a supportability pointer for algal biofuels. It is essential to evaluate field, glades and waterfront waters that could be moved from crude material creation offices. The pertinent measurements are:

- National or local zone of glades and forests committed to the field of dairy cattle; in any case, the absence of information at first glance utilized for munching animals.
- The number of cows bolstered in fields and forests.
- Performance pastures in view of biomass or scavenge surface.

A less immediate pointer of the measure of dairy cattle or biomass would be secured by fields and bushes. Alternate pointers have recommended supportability endeavors for the recuperation of defiled regions. Some of these arc summed up at the wedding function and Crawford-Brown (2007) and be suitable where the generation of algae biofuels is found in deserted fields.

GREEN HOUSE GAS EMISSION

The capacity to diminish ozone depleting substance outflows is one reason for the improvement of biofuels. The essential relief is that the burning carbon discharges of relating biofuels obtained in photosynthesis. All things considered, fuel and co-discharges of ozone depleting substance emissions are zero carbon and other nursery gasses transmitted amid machining. In this area, the aftereffects of the life cycle life cycle evaluation considers (LCA) are basically explored.

Life-Cycle GHG Emissions of Algal Biofuels

It is normal that ozone harming substance discharges from the fundamental algae biofuels are identified with the vitality utilization in the handling chain. The utilization of ozone depleting substance vitality is entangled by the inconstancy of the carbon over-burden of different types of vitality, particularly power. Joined States Generation Gas Greenhouse Energy Effect Direct Emissions Directing This is 606 grams CO2 equal for every kilowatt hour.

LCA comes about for net ozone harming substance emissions for the generation of algae biofuel extend from the net negative esteem (ie a carbon tank) generously higher than the positive fuel oil esteems. As on account of vitality utilize, CO2 changeability factors are the wellspring of supplements, profitability and process execution, and acknowledge related for coproducts. For instance, Sander and Murthy (2010) expect that remaining algal biomass replaces ethanol corn plants.

Substitution of corn oil extricated from ozone harming substance algae crude material for ethanol brings about a negative carbon adjust. For reference, coordinate carbon burning is around 2.7 kg CO2 proportionate per liter of fuel.

As said above, transformation of algae grass trim structures may require field changes somewhere else. Be that as it may, vulnerabilities are too extensive to evaluate this likelihood or to figure net ozone depleting substance discharges in these theories.

Albeit numerous horticultural procedures discharge ozone depleting substance copolymers, for example, nitrogen dioxide (N2O) and methane, these outflows have not been experimentally settled as noteworthy for algae development. N2O could be discharged from development frameworks, and these emissions ought to be measured later on for development conditions that could advance or transmit methane N2O. A solitary animal types consider measures N2O emissions from algae development under research facility conditions. In this examination Nannochloropsis nitrate ascends as a wellspring of nitrogen, high N2Os were seen in a nitrogen head space (photoreceptor reproduction) in times of haziness, yet N2O was low amid light periods. On the other hand, when the air space comprising of air (open tuning reproduction), N2O was unimportant. Denitrifying microscopic organisms were available. Nitrites and nitrates are utilized as a part of microbial denitrification decrease with N2O era and, at last, nitrogen gas. Anaerobic airs are essential for change, however high denitrification rates are watched for which oxygen is accessible in accessible option (Kleiner, 1974). In waterways, lakes, lakes, and estuaries, N2O generation is corresponded with nitrate fixations in water (Stadmark and Leonardson, 2005). The denitrification rate relies upon the hidden soil and the porousness of the covering.

It is vague whether anaerobic denitrification is the main potential method for producing N2O in algae culture frameworks. Metaneogenesis can happen in freshwater and marine dregs, soaked water, bogs, swamps where oxygen is low. These conditions can win in a few lakes with significant biomass or other natural issue in the silt. Methane is discharged when natural acids, alcohols, celluloses, hemicelluloses, and proteins are corrupted. Methane creation is identified with water temperature and is expanded at impartial pH. Metaneogenesis is stifled by nitrogen aggravates that microorganisms can be utilized as an electron acceptor, amongst nitrates and nitrites, yet these mixes can without much of a stretch be lessened to devastated oxygen conditions. Methane and denitrification could be enhanced if the collect comes up short. Amid calamitous disappointment of culture, thick algae dormant algae biofuels can wind up noticeably anaerobic and emanate an assortment of nitrogenous or mixes containing unpredictable sulfur and methane. Notwithstanding, culture disappointments ought to be of brief span and once in a while, if biofuel organizations need to keep up a net revenue.

Opportunity alleviation

An extra potential to alleviate GHG utilizing low vitality hotspots for carbon change and carbon power substitution. For instance, the advantages of the bioelectricity carbon generator are higher in zones where the lattice depends on petroleum derivatives. The yields and properties of the different co-generation choices are minimal known. The capability of N2O and methane could be lessened via watchful blending and legitimate administration of algae crops.

Sustainability Indicators

A decent ozone depleting substance manageability discharge marker is the measure of CO2 transmitted per unit of vitality created, picked as a pointer of ozone depleting substance outflows and usually used to talk about ozone depleting substance emissions.

LOCAL TIME

Potential environmental effects

The presentation of a lot of water in dry or semi-dry conditions could adjust neighborhood climatic zones by expanding stickiness and decrease outrageous temperatures. Ecotourism ponders give valuable data. Waterways made by repositories could influence the dissipation rates of the encompassing scene, causing changes in plant cover and assorted variety of land species. Huge dams can influence time and precipitation, especially in the Mediterranean and semi-dry atmospheres.

Sustainability pointers

Supportability markers for potential neighborhood environmental change are the relative stickiness patterns and patterns in the measurable circulation of temperature.

Information and Data Violations

In spite of the fact that it can be determined in parallel with the presentation of extensive dams in dry areas, dimensional inconstancy, geology and rising creation techniques, for example, the algae development industry require additionally research to comprehend and address the effects related with neighborhood environmental change.

QUALITY OF AIR

Potential Environmental Effects

The air quality effects of algae biofuel generation rely upon the plan of the framework. Different issues increment air quality in conjunction with the different periods of the biofuel chain. In this manner, this area is composed along ventures along the creation lines. The extensive variety of conceivable outcomes for creating algae biofuels and an extensive variety of powers closes in an extensive variety of conceivable climatic discharges.

This segment shows the exceptional air quality for the creation of algae biofuel discharges and does not represent non-renewable energy sources utilized for non-renewable energy sources or copying gear that can be utilized for the generation of manures or pesticides. The motivation behind this part is to consider discharges for the generation of algae biofuels, with the goal that proper pointer are distinguished. Be that as it may, outflows from petroleum products as algae fuel generation ought to be considered in any effect on the impacts of air nature of different biofuel algae outlines. Also, we don't know how much algae biofuels will extend and how air quality can change with expanding scale.

Open-Pond Cultivation

The Committee is ignorant of conceivable environmental poison discharges from lakes bolstering on biofuels in writing. Under ordinary working conditions in pools, crops are relied upon to be oxygen consuming and low unstable natural mixes (VOCs). In any case, smaller scale algae and large scale algae develop in characteristic marine situations are known to be essential wellsprings of VOC, amongst isoprene and monoterpenes. Texas An and M University analysts are as of now assessing and measuring VOCs from an extensive variety of algae and developing and freshwater pasta. Three of the species inspected are developed for open biofuels, open and shut repositories with photobookers CO2 tests got from crops in wastewater. In preparatory outcomes, unpredictable natural mixes were recognized as 45 mixes.

Other airborne emissions are relied upon to be transmitted straightforwardly or in the environment made by sulfur dioxide antecedents (SO2), nitrogen oxides (NOx), NH3 and VOC. Pressurized canned products may incorporate algae and supplements and in addition an extensive variety of microalgae mixes, including poisons. Microalgae are known characteristic wellsprings of regular marine airborne sulfate.

Numerous musty algae deliver optional metabolites, however these algae are probably not going to be chosen for substantial scale generation. Scents delivered amid vigorous development as auxiliary metabolites. Different smelly mixes are related with the decay of algae under anaerobic conditions in which the microscopic organisms corrupt the natural issue and create hydrogen sulfide and NH3, both with a solid scent. In open holders for algae culture, anaerobic conditions are limited. Photovoltaic emissions in open tanks will be lower if undesirable vaporous items and foul chemicals previously permitting gas trade with the outside condition are dispensed with.

Drying

Drying forms used to deliver coarse and fine particles. Groupings of suspended particles rely upon the advancements utilized; For instance, drying belts and transport frameworks prompt an expansion in nearby detached sun oriented drying emissions. On the off chance that the outflows go past the establishment will rely upon the control level. Particles can likewise be an expert hazard even in shut situations. In dusty situations, clean might be a danger of blast. Poor drying techniques can likewise prompt the disintegration of biomass and the arrival of VOCs, amines, methane and different mixes.

Extraction

Numerous treatment strategies incorporate the proposed algae biofuel cells or other lipid aggravates that utilization natural solvents. Extraction with natural chemicals, as required, some create dissolvable discharges, and the sums discharged rely upon the innovation utilized. The most widely recognized dissolvable that is transparently talked about by makers is hexane. In a natural evaluation, Sapphire Energy, Inc., one of the biofuel algae, said under 50 ppm hexa stays strong algae after a burning recuperation process. This leftover hex loses the strong algae in the air amid the algal biorefinery sanitization process (IABR). Alluring properties of these solvents are reasonable, tough, low polar structure poisonous quality and poor lipid segment extractor cells. The hexane has been utilized as a dissolvable for oleaginous plants for the creation of biodiesel versus hexane tanks can limit the arrival of algae disolventes. ibles includes extraction of lipids or other natural solvents made out of cells. Natural extraction items, fundamentally, some deliver dissolvable emissions, and the sums discharged rely upon the innovation utilized. The most widely recognized dissolvable is transparently examined by makers is hexane. In a natural appraisal, Sapphire Energy, Inc., one of the biofuel algae, said under 50 ppm hexa stays strong algae after a burning recuperation process. This is a strong waste release of hexagonal algae in the environment amid transport to the decontamination procedure of seawater oil purifier (IABR incorporated). Alluring properties of these solvents are minimal effort, strong, low-lethal, nonpolar and poor structure. The test part does not have a lipid extractor utilized as dissolvable vegetable oils to make biodiesel tanks or to follow directions that can limit the dissolvable discharge exposition.

Pyrolysis

They are trying innovation to change over aggregate biomass into fluid energizes. These procedures can have various voices with various nourishment and air emissions caused by the generation of esterified or green diesel. Pyrolysis of biomass produces three vitality items, for example, strong (coal), fluid (bio-oil and gas) in various extents relying upon temperature, weight, habitation time and different components. Gasses are reused to control the framework and in this manner don't straightforwardly add to outflows, with the exception of on account of criminal emissions that could get away from the framework. Warming of pyrolysis units could contribute a little measure of NOx and carbon monoxide (CO). It might be fundamental more vitality is likely nourished by gaseous petrol to dry algal biomass adequately before pyrolysis. Particulate emissions, corrosive gasses and pyrolysis hydrocarbon discharges are not portrayed by writing. Bio-oil created by entire cell pyrolysis requires promote change for the generation of transport energizes. The change of should be possible with a different hydro treating step or a comparable pyrolysis process and hydro-incorporated hydroprocessing forms. Regardless, hydrogen is required. Hydrogen generation delivers low levels of NOx and produces a CO2 stream that could be utilized to sustain algae crops.

Anaerobic Digestion

Anaerobic treatment refinement plants deliver absorption of biofuel algae it is noticed that NH3 is available in biogas by anaerobic processing at focuses up to 450 ppm. The grouping of NH {3} in biogas relies upon the nitrogen substance of the specific bolster material. The principal works by a specialist found that anaerobic algae absorption decided N2 and NH3 centralizations of the request of 4 to 5 percent (normal in volume) of aggregate gas creation. NH3 would not be discharged for air around the office because of the want to reuse the supplements required for development.

Life-Cycle Assessment

Emissions of atmospheric pollutants should be valued biofuel algae in cycle and compared with petroleum fuels and other alternatives. A study (2009) and the data illustrate the importance of this assessment. It has been observed that while the use of gas and biofuels (corn ethanol and cellulose ethanol) releases similar amounts of VOC production of PM, NO_x, SO_x and NH_3 are significantly different oil and fuel. Biofuels emit a greater amount of fuels derived from VOC, NO_x, NH_3 and PM 2.5 petroleum. The Committee has no knowledge of atmospheric pollutants such as ACV for algal biofuels. This analysis is critical to assess whether the production and use of biofuel lead to improve air quality with respect to fossil fuels and provide information on the chain phases significantly contribute to atmospheric pollutants.

Potential Social Acceptability Effects

Algae produce a number of sprays and secondary or harmful metabolites for humans, some of which may be harmful (eg, odorous). Also, some chain processes such as extraction and drying, or particles may emit solvents that could affect local air quality if they are not contained. If a biofuel plant near human populations measures to restrict or limit the release of any product that adversely affects air quality at the local level or is likely to be a public health risk. The health costs of certain types of atmospheric emissions have been discussed in literature. Depending on the quantity of these products and the proximity of inhabited areas for crop production, the reduction in air quality and the risks to health and quality of life may affect the Settle perceived process and allow permits. If public opinion is not aware of these potential effects before sitting down and allowing for installation, there is the risk of unwanted compounds can not be considered once the structure is built. If this is the case, disputes or protests can slow down or halt operations, resulting in financial losses for the developer and negative attention for the industry in general.

Opportunities for Mitigation

The photobioreactor could have reduced the effects of open water quality systems. However, a section with atmospheric emissions associated with the production of materials and the functioning of the bioreactor system to assess whether the photobioreactor is a small or negligible impact is necessary for the air quality. Although passive processes (eg solar drying) reduce the impact of air quality on active processes that generate dust rates or increase volatility, they are not large-scale practical solutions. Facilities located at a distance from human habitats and species of ecological interest mitigate the possible adverse effects of air pollution in human beings.

Sustainability indicators

Appropriate air quality metrics depend on the processes used in the production of seaweed biofuels. Concentrations should be measured or molded into appropriate regulatory levels or potential health or discomfort effects for human scales. These may include:

- For open tank systems, VOC concentrations and odorous secondary metabolites.
- For active drying processes, concentrations of particles suspended in the air.

- For extraction processes, concentrations in solvent air used.
- For pyrolysis, particulate matter, hydrocarbons and acid gases.

Information and Data Gaps

Measurement of open atmosphere emissions to provide information on occupational exposure estimates and other environmental factors that can be compared with human health thresholds or environmental storage tanks. The lack of information and data including the relationship between some drying technologies, types and concentrations of released particles, solvent emissions during mining potential NH_3 concentrations in the air for anaerobic digestion and potentially released pyrolysis products. This information will be presented when bioraffinerie search for air quality permits.

AQUATIC BIODIVERSITY AND SPECIES INVASIVENESS

The species's invasiveness is the exclusive concern of biofuels produced by algae and vascular plants. In addition, the change in soil use or the alteration of landslides for the production of raw material biofuel algae can affect biodiversity. The effects of many biofuels raw materials on biodiversity and the mechanisms leading to these effects begin to be understood. However, existing studies focus mainly on ground and ground ecosystems for biofuels instead of algae and raw material systems.

Algae distribution in natural aquatic environments

Many eukaryotic cyanobacteria and microalgae are cosmopolitan in their spatial distribution (biogeographic) and therefore cannot be invasive if released into regions of its wide range of habitats. However, not necessarily in all places where requirements are met habitats (eg, pH, salinity, temperature, humidity and time) so that its distribution is often mosaic. Other algae can be endemic in particular regions, for example, some cyanobacteria in Swedish lakes and particular marine species. Endemic species may become invasive if transported to other places, but these species may also exist in small quantities elsewhere, although there are no recorded

The algae may be larger than the site due to lack of sampling on some continents (in particular benthic habitats) and the lack of detection of low-density bodies. Coastal marine macro algatends to be less cosmopolitan in the distribution of phytoplankton and micro algae cyanobacteria space. Macro-algae have a narrower temperature, light, and preferences of the nutrient substrate.

The wide range of processes that could lead through open micro algae water can also contribute to dispersion and therefore to a wider distribution. Many species of algae can be transported by air. Algae vectors include aquatic insects, dragonflies, and bees, aquatic mammals living in the beetle, such as raccoons, mink and moss. Algae main vectors are birds. In a study of 16 species of aquatic birds, 86 species of algae, 25 species and 25 species of feathers found. Most algae survived surface water for four hours, but most did not survive more than eight hours. Some species of algae may seem rare. Whitford (1983) explains that freshwater algae species may seem strange for several reasons (for example, rare historical collections, long-spore species that easily germinate and very specific species in their habitat requirements), very few species Fresh water are really rare. This suggests that some species of rare algae can be moved by invasive algae used to produce biofuels raw materials.

Releases of Algae to Natural Environments

The probability of releasing an open pond would be related to the surface and free space (ie, the distance between the normal water level and the top of the culture tank), the speed and direction of the prevailing wind frequency and a quantity of Rainfall splashes), water body distance, probability and intensity of potential vector visit, such as water birds and mammals and absolute abundance (cells per ml) of species that could be released into the environment. Humidity affects the survival of unicellular algae. Survival rates vary among seaweed groups. In a study, climatic conditions such as temperature, relative humidity, precipitation, wind speed, sunlight influenced release and vertical seaweed transport.

Spreading into distant sites can occur through the air, through the water, and the animals or boats or vectors. The wide ranges of vectors that can eliminate algae pools open aquatic insects, including dragonflies, birds, and raccoons, among others. Closed photobioreactors have a much lower risk of seaweed release and transport. Open or closed system collection operations could be an important potential route for the loss of microalgae in the surrounding environment.

Potential Environmental Effects

Distant distances can occur through the air, through water, and animals or boats or vectors. The wide ranges of vectors that can remove algae pools open aquatic insects, including dragonflies, birds,and raccoons, among others. Closed photobioreactors have a much lower risk of seaweed release and transport. Open or closed system collection operations could be an important potential route for the loss of microalgae in the surrounding environment.

Opportunities for Mitigation

Invasive species of algae that can cause harm or flowers cannot be selected for large-scale fuel cultivation. The invasion varies in different natural environments and may require specific site qualifications to reduce the risk of invasion. In addition, conditions are intolerant species in natural waters (eg salinity) near the installation of biofuels can be selected to minimize the risk of invasion if released.

Landscape design can also be considered to limit the impact on the algae unlocking system. Systems away from watercourses and wetlands where ponds can thrive could reduce or minimize the likelihood of flowering species being released. In light of the factors influencing the probability of release and the abundance of its release, mitigation measures may include wind shields and mechanisms to discourage vector organisms.

Sustainability Indicators

Eco-sustainable community with low environmental impact diversity includes sustainable ecological communities that include aquatic diversity metrics and algae invasion. A category of such metrics would be diagnostic capabilities for invasiveness. Qualitative metrics related to invasiveness, but not necessarily diagnostic, include:

- Rapid growth in natural environments.
- Extensive habitat tolerances, for example, temperature tolerances, light and nutrients.

- Resistance to parasites and herbivores.
- Aggressive for resources, such as light, nutrients, and space competition.

It advocates more direct measures of aquatic biodiversity regarding the sustainability of biofuels. (2011) are relevant here:

- The presence of taxa of particular interest. These may include rare fish, aquatic invertebrates, or macrophytes.
- Taxa habitat area of particular interest to aquatic organisms that could translate to the length of the current field for taxa and invasion of algae. A category of such metrics would be diagnostic functionality for invasiveness. Qualitative metrics related to invasiveness, but not necessarily diagnostic, include:
- Rapid growth in natural environments.
- Extensive habitat tolerances, for example, temperature tolerances, light, and nutrients.
- Resistance to parasites and herbivores.
- Aggressive for resources, such as light, nutrients, and space competition.

It advocates more direct measures of aquatic biodiversity regarding the sustainability of biofuels. (2011) are relevant here:

- The presence of taxa of particular interest. These may include rare fish, aquatic invertebrates, or macrophytes.
- Taxa habitat area of particular interest to aquatic organisms that could translate to the length of the current field for taxa of interest.

TERRESTRIAL BIODIVERSITY

Landscape Pattern of Development

Potential Environmental Effects

The scene display for transformation for any new foundation could influence earthbound species and assorted variety of the group through no less than three unique instruments additionally apply to the creation of algae biofuels or other vitality generation:

- Move the world's vegetation and untamed life living space into the establishment range and supplant it with a lake or photobioreactor containing a monoculture or ocean growth.
- Reduction of neighborhood natural life territory beneath the edge required for this species.
- The fracture of natural life territories for the most hard to discover couples or relocation halls are interfered.

The removal of local and single vertebrate vegetation is for the most part constrained to the territory of the plant, however a few species are delicate to human foundation and tend to go past the points of confinement of the plant, for instance, the gathering of flammable gas framework.

Substantial foundations, particularly for most establishments, some living vertebrates could section expansive range. The fracture is controlled by the surface no less than one plant living space or noteworthy sprinter's module. Fracture is a measure of the boundary (fix edge length) and fix environment region. Subsequently, a direct framework tends to part into numerous situations in a way that is nearer to the square. In any case, this arrangement is more down to earth for framework upkeep, so they are not viewed as huge direct structures. Other conceivable activities are: recurrence fracture of the scene involved by a given living space, number or thickness of environment fixes in a given region (more fixes mean further discontinuity) and the level of confinement or living space disconnection.

Notwithstanding when the living space is divided human issue related with foundation and could decrease the living space zone past the base levels required for a few animal types. They have discovered that few examinations have taken a gander at dynamic conduct or individuals related with substantial bordering natural surroundings zones, including little grimy soils or modified inadmissible, (for example, the advancement of biofuel algae or oil and gas improvement). An exemption is a hypothetical report by American Badger at the Oil Production site contemplated the impacts of an assortment of other expensive shrimp in Oklahoma. The basic regions for unsettling influences rely upon the loan cost, living space sort, natural reasonableness, and kind of framework.

The effect on verdure may change broadly contingent upon the particular site picked and the transcendent utilization of gauge soil flow and no development in algae and refined biofuel refineries. As indicated by a resaerch conceivably reasonable biofuels incorporate soil sorts, including 42 for each penny bushes or brambles, 19 for every penny of the cultivable land, 14 for every penny of evergreen timberlands, 10 for every penny of herbs, 8 for every percent. The most ideal soil and water in the Gulf Coast locale are essential conditions. Woodland natural surroundings in the US are across the board, however are debilitated by changes in arrive utilize designs; Many types of fowls rely upon this living space are falling. The advancement of huge extends of tropical bushes of 181,000 square kilometers could quicken the decay.

Opportunities for Mitigation

The presence and abundance of wildlife must be evaluated before construction, such as plants that are subject to landscape design environmental assessment could minimize the possible effects on biodiversity. A research suggested that integrating design considerations for bioenergy could prevent or minimize the adverse effects on terrestrial biodiversities, such as maintaining corridors for the wildlife movement. When planning the size of individual basins, their density in the landscape and the associated production structures, managers should consider the potential environmental impact on biodiversity.

Wildlife Alcohol

Potential effects on the environment

Open algae lakes can be water hotspots for untamed life that might be helpful in parched or destructive conditions in the event that they are poisonous to a few animal categories. The hazard that creatures are presented to saltiness or chemicals in water tanks for the generation of algae is obscure and has unfriendly impacts from presentation to water or skin introduction.

The harmfulness of introduction to salt is conceivable. This happens when chloride salt gathers in blood at dangerous levels and, on account of winged animals, at over the top rates for discharge of salt organs. For instance, sodium touchiness harmfulness has been seen in southeastern New Mexico lakes. In any case, it is improbable that water for algae development is hypersaline. Shorebird species are specific living beings to suit elevated amounts of salt. The deadly and sublethal convergences of saltiness for a few animal types are condensed in a report from the US Department of Energy (1998), with poisonous quality edge esteems for ducks extending from 9 to 20 sections for each thousand saltiness a large portion of the ocean water at 35 ppt).

Numerous synthetic and behavioral variables could influence the introduction of natural life to salt and different chemicals in open tin frameworks. For instance, the advancement of simulated water in forsake conditions is once in a while an essential wellspring of water for neighborhood winged creature populaces yet might be less critical for some transient species or creatures that may have a high level of devotion to water sources particular. Water is probably not going to be uncommon in the scene and natural life will have numerous choices for water. Thick kelp lakes may not be as appealing to natural life as the purest water, yet this speculation has not been demonstrated. In like manner, the thick impact of algae in fascinating lakes in untamed life is obscure. Vanishing lakes for wastewater from oil fields Bird presentations give off an impression of being verbose, concurring with transient conduct. To consider potential introduction to untamed life in harmful biofuel development waters and their potential impacts may be like the dissipation of lakes and oil reflux agglomerations.

In the western piece of San Joaquin Valley, California, dissipation tanks ranches where there are different choices for wastewater seepage have been produced. Flying creatures utilize vanishing lakes to rest, searching and settling. An examination on the utilization of limestone fowls in the Central Valley of California has discovered that horticultural vanishing lakes are exceptionally alluring to these winged animals. Another examination found that water does not dissipate chose lakes or seepage or cleaning plants (which may resemble some biofuel lakes) and most loved conduit

For horticultural dissipation lakes, the fundamental worry of fauna was the grouping of selenium, its natural changes and gathering of farming vanishing crevices, the primary worry of fauna was the centralization of selenium, its ecological changes, and the earth considered. In the event that selenium can speak to critical presentation algae, it relies upon the accessibility of selenium in the source water and the hidden soil, if the water channels the can. A few scientists recommend that oceanic winged creatures presented to the horticultural division in the water dissipation bowls could be in danger of uranium harmfulness. The uranium aggregation in lake dregs was somewhat ascribed to the deterioration of algae. The progression of arsenic has additionally been considered as a potential concern.

In another conceivably comparative illustration, it has been watched that flying creatures and bats, creatures of land and water, reptiles, little well evolved creatures, diversion species and creepy crawlies are pulled in by substantial dissipation holes of 0.4 to 2 sections of land of US squander transfer office squander regions in the USA. Dead fowls of these lakes are by and large credited to oil, however sodium danger and surfactants have been ensnared sometimes.

Pulling in algae lakes could be a major ordeal on the off chance that they contain lethal chemicals or pathogenic operators at destructive levels. Unplanned dead winged animals and fish have impacted social (and in this way maintainable) hydroelectric and wind advancement, individually. Versatile administration can assume a part in moderating the negative impacts of untamed life through drinking introduction.

Opportunities for Mitigation

The Committee doesn't know about the reports that natural life is a worry for existing plants to develop open algae open. Due to the number and size of the structures, it is important to check the potential convergences of poisonous substances in water and presentation to drinking water fauna to guarantee that it is not a worry. A portion of the lessenings utilized for oil tanks in landfills cover the surface with plastic balls to make them less alluring to lakes are not choices for photosynthetic fuel sources. Similarly, relief systems utilized as a part of horticultural dissipation lakes, for example, rising lakes or keeping up profound water levels, decreasing winged animal wellness propensities, are not down to earth for algae development that requires shallow lakes. As open tanks are controlled by science, harmfulness limits for these substances help decide when gathered water is to be expelled and reestablished.

Sustainability Indicators

Metric biodiversity earthly for the maintainability of biofuels that suggest by a scientist and applicable for issues identified with the scene improvement show are:

- The nearness of concern taxa (nearness).
- Habitat territory of specific taxa intrigue (hectare).

For natural life exposures for saltiness and contaminants in drinking water, maintainability pointers include:

- The dosage got from fauna (coordinate estimation).
- various vertebrate passings from drinking algae lakes every year (coordinate estimation).
- Toxic focuses, poisons or saline culture medium (less immediate estimation).
- Abundance of vertebrates that drink from open lakes every year (less immediate measure

Information and Data Gaps

Algae biofuel plant advancement models in connection to regular hallways have not been examined in light of the fact that the spots for future improvement are indeterminate. It is important to comprehend the spatial example and scene of these advancements to recreate the consequences for natural life populaces. As algae culture is extended in number and size, it is important to assess the potential for drinking water locales. On the off chance that you watch the utilization of untamed life, at that point you should quantify the groupings of harmful substances in the waters of the wellspring and the rising waters to guarantee there is no peril to the wellbeing of natural life. On the other hand, measures can be taken to demoralize drinking water.

GENETICALLY ENGINEERED ORGANISMS: ENVIRONMENTAL EFFECT

Potential Environmental Effects

The ecological manageability of hereditarily adjusted crude materials for bioenergy and the potential ramifications of the enactment on practical advancement in the business have been viewed as before, yet the accentuation has been ashore based product changes in algae. Some biofuel algae organizations, for example, Algenol and Synthetic Genomics are leading examination on hereditarily altered living beings for the generation of algae biofuels. In a speculative situation, the most noticeably bad, hereditarily adjusted algae that have been brought into regular habitats can persevere and turn out to be abundant to the point that they make unsafe algal sprouts. Unmistakably, any antagonistic impacts of hereditarily changed algae discharged, if watched, would undermine the reasonable improvement of algae biofuel innovations. The evaluation of the potential impacts of hereditarily adjusted algae will be an unpredictable undertaking and, given the decent variety of life forms, the scope of building capacities and the scope of situations that are possibly controlled

Distinctive attributes of algal biofuels can be adjusted with hereditary building techniques. The majority of them are proposed to expand biomass or oil efficiency, albeit some might be intended to limit survival or propagation after discharge. More prominent profitability could include objectives, for example, expanded lipid content as a forerunner to biodiesel (which could prompt development in nitrogen-insufficient nitrogen-rich or sans silicon cells), the presentation of organic pathways for coordinate fuel creation requires negligible treatment preceding Distribution and utilize, altering cells to discharge crude material or fuel specifically into the way of life medium, adjusting sugar digestion in cells (expanding glucose stockpiling, lessening starch debasement), expanding stress resistance factors, (for example, salt, light , PH, temperature, glyphosate) disinfectants. Some of these parts of designing and purposeful and unintended going with characteristics could concern both the wellness of algae for the creation of biofuels, for example, their survival and physiology when discharged into normal frameworks. Phenotypic changes that could prompt noteworthy natural impacts of the discharged life forms incorporate those that convert into

Increments in physiological resistance or adjust the utilization of the substrate or to alter the geographic dispersion of the species. Indicators of potential unfriendly impacts of hereditarily adjusted algae are probably going to discharge, the wealth of living beings discharged (indicator setting), the survival and wellness rate, reproducing rate, separations to removed locales, association with different creatures, Probability of hereditary trades and the probability of a negative impact. The new elements are conceivably ready to impact these elements, yet few of these connections are incorporated. The phone thickness in the way of life medium could be influenced by the designing attributes. The adequacy and recurrence of outflows can decide if the discharge prompts the self-maintained (built up) populace. The survival rate of hereditarily altered microalgae or cyanobacteria is dictated by the blend of the personality of the species, (s) change (s) quality (s) and around which it is discharged. High lipid algae are probably going to be more alluring to predators. A few scientists recommend that most GMOs have a littler limit in beneficiaries of unmodified living being conditions. Algae can be crossed or intended to have high development rates under particular culture, and some of them may have high rates of development in particular characteristic conditions. The new attributes gave by hereditarily adjusted algae decide if and how group connections could be changed.

The three sorts of level exchange preparing are free desoxyribonucleic corrosive (DNA), conjugation and transduction. Exchange of qualities between living beings is normal in a few animal groups. Around 1 to 20 percent of the bacterial genome comprises basically of DNA obtained as of late by different prokaryotes, yet in addition eukaryotes, for example, methimazole. Hereditary trade amongst prokaryotes and eukaryotes is not all around considered past particular communications in sets, for example, the exchange of Agrobacterium T-DNA species to plant cells. Some early developmental exchange exchanges have been recorded. For instance, a quality for fructose bisphosphate aldolase coordinated by plastids of red algae and some Prochlorococcus Synechococcuspecies exchanged. Another examination, for instance, indicated confirm for even exchange of a self-grafting intron and homing from cyanobacteria (Calothrix) to the chloroplast Euglena myxocylindracea genome. Exchanging qualities between various living beings are conceivable, yet uncommon. There is prove that nucleic acids encoding polysaccharide proteins have been exchanged from an ascoglossane to ocean growth a snail devours marine algae. On a level plane exchanged qualities can encode chose elements, for example, imperviousness to anti-infection agents and metabolic pathogenic proteins. Level quality exchange relies upon the thickness of creatures with which trade is conceivable.

The primary antagonistic impacts of any algae, hereditarily altered or not, could incorporate biological and wellbeing impacts of poisons, natural blossoms, and species substitution. These potential impacts are talked about somewhere else in this report. Potential markers scattered imperviousness to anti-toxins likewise be a worry, however the animal categories that contain these markers presumably can't be utilized for biofuel generation on a business scale. The creation of hereditarily adjusted algae poisons is far-fetched in light of the fact that the maker of lethal strains or strains liable to be intended to kill poison qualities would be maintained a strategic distance from. A hereditarily altered strain may have a lower danger of a negative effect than a characteristic strain that did not have such changes. The potential classifications of hereditarily altered life forms that have been featured by analysts ought to have been considered in the assessments of algae discharged. This incorporates the making of new or more compelling pathogenic specialists that influence non-target species that bother groups and biotic environments, decrease biodiversity and hereditary assorted variety of the species.

Most ways to deal with chance evaluation recommend that commonality with hereditarily altered creatures is a vital indicator of hazard. In particular, hereditarily designed life forms that have a background marked by safe use in applications like adjusted ones (e.g. comparable densities comparable biological systems) applications don't debilitate the natural maintainability of algal biofuels. Moreover, microorganisms don't create from various life forms of inception, however are made by firmly related living beings are more averse to have new capacities and cause unfavorable impacts.

Social Acceptability of Genetically Engineered Algae

A few open and political concerns have been expected with the Development of hereditarily designed algae. One is significant concern is about the survival of such species in the open condition. The utilization of hereditarily built microorganism in sustenance preparing industry is additionally an issue of worry among open. Be that as it may, the social adequacy of these species is additionally reliant on their application, for instance, medicinal applications. On the off chance that the researchers could demonstrate their use and advantages for medicinal related applications that could be adequate, in any case, similar to application in sustenance industry and agribusiness are worried, there are more open level headed discussions about the moral, wellbeing and danger associated with their utilization when contrasted with their advantages. It is to a greater degree a matter of trust.

Opportunities for Mitigation

Control of hereditarily adjusted algae could be alluring as a safety measure against the obscure impacts and worries of society. Physical limitation of algae discharged will be troublesome or unthinkable. Physical limitation arrangements, as proposed for vascular plants (e.g. fencing and fringe plants) are inadequate against the algae discharged. Control choices may incorporate the utilization of species that require salt water in freshwater situations or those with a supplement necessity is not fulfilled outside the photobioreactors or lake. It was proposed to utilize it relying upon nature to improve the probability that the group acknowledges hereditarily altered "atomic switches" societies. In like manner, some changed styles may diminish wellness in a regular habitat. For instance, lessened light gathering recieving wires increment development in lakes as well as diminish the capacity to rival local types of algae from light in characteristic waters. Also, the disposal of bicarbonate pumps, which builds reasonableness in the high-CO2 condition, could decrease the aggressive capacity to retain inorganic carbon in regular waters. In the event that algae are utilized, they will be compelling against non-target life forms. Eliminator qualities that reason dead kick the bucket cells could be created such that erased qualities in photobioreactors yet discouraged in indigenous habitats. In the event that you utilize algae strains that can't deliver poisons, the potential hazard is limited. In the event that hereditarily changed life forms are discharged, observing ought to be made for the impacts of particular mixes between the life form and nature can be better caught on.

Sustainability Indicators

Maintainability pointers for hereditarily adjusted algae would by and large be similar markers for local algae. That is, if the impacts of intrigue incorporate biodiversity or water quality, the proper measurements are depicted in these parts. Nonetheless, manageability destinations for hereditarily changed algae would likely incorporate two different issues:

- Minimize the spread of hereditarily changed algae.
- Establish techniques to decide if a watched impact was caused by hereditarily changed algae.

The wealth of hereditarily changed algae in water discharged could be measured by particular trial of the species if the species were not local. What's more, some adjusted segments, for example, changed recieving wires might be minutely recognized and in this way quantifiable water. They may likewise be discernible by specific DNA arrangements. Likewise, pointers could be added to algae to consider simple estimation of particular media

Information and Data Gaps

The environmental dangers of the arrival of hereditarily adjusted creatures must be precisely assessed before being utilized as a part of the generation of biofuels on a business scale. On the off chance that there are conceivable situations to which hereditarily altered algae or living beings obtain qualities from hereditarily changed algae could develop to levels, for example, damage to people or nature, they should somehow be inspected. More data on the conceivable connection between qualities that are behavioral alteration objectives and cyanobacteria or miniaturized scale algae that could adjust the rates of discharge, survival, development, transport, hereditary trade and impacts on human or biological wellbeing is important. In the United States, they led a little research on the conduct of hereditarily adjusted algae in open lakes, somewhat in light of the fact that the APE revealing rules can cause delays for scientists. Data is required on the social worthiness of the utilization of hereditarily changed algae for biofuels, especially in open frameworks.

WASTE-PRODUCTS

Potential Environmental Effects

Strong squanders from algae biofuel fabricating forms will probably be produced as an anaerobic digestion related mud than unstable natural acids are changed over into methane and CO2; Methane is utilized as a fuel supplement for the procedure. Anaerobic absorption is regularly trailed by oxygen consuming assimilation to change over disintegrated solids to sedimentation of slime packed in expansive vigorous colonizers. These frameworks have been monetarily worked for quite a long time and will undoubtedly be incorporated into algae biofuel creation forms and processed fixations. An examination announced a normal $NH4^+$ mud in the scope of 1600 to 1850 mg for every liter for anaerobic absorption of algae, which is similar to a portion of the most elevated esteems detailed for cochineal transfer. Another wellspring of strong waste is engineered plastic covering utilized open or shut bioreactor tanks that should be occasionally sorted out.

Life, as indicated by Jim Sears, who led the "Specialized Standards Committee" for the algal biomass association, "there are such a significant number of procedures engaged with creating biofuel algae as there are organizations." Therefore, if squander era is an issue it can't be known until economically executed operations and can decide sytheses. Amplifying reusing would decrease the requirement for squander transfer.

Opportunity relief

Supplement reusing is the conspicuous waste alleviation creation. Algenol aalgae biofuel organization intends to reuse seawater squander for development (under 6 liters of seawater squander per liter of ethanol delivered if the photobioreactor last over 6 years).

On the off chance that processed muds happen, purifiers, urban waste for the most part spread rich supplement waste to assigned, arrive, with the advantage of conditions and sustaining the dirt. It doesn't plan to deliver sewage ooze fuel for ocean growth wastewater is fundamentally unique. The procedure of approval of the National Pollutant Discharging Waste Regulation directs ooze release; most states have an approval procedure under this government program. The muck structure is checked to guarantee consistence with the allow.

Sustainability Indicators

Numerous applicable pointers for maintainability deposits are portrayed somewhere else, for instance, measuring reusing of supplements and saltiness of groundwater. On the off chance that the saline water in groundwater is infused, maintainability markers may likewise incorporate the yearly volume infused volume every year.

Information and Data Violations

Data on sorts and speeds of waste era for most seawater generation forms is biofuel.

POTENTIAL EFFECTS

Potential consequences for the earth

Algae Toxins

It is improbable that known strains that create poisons in the generation frameworks of algae utilizing biofuels. Truth be told, numerous species have a nourishment review status or are utilized as encourage in aquaculture. In any case, some very much considered species may really deliver beforehand obscure poisons. Cases of these incorporate the current eugenia revelation of poisons and free radical poisons. Likewise, contaminating algae and cyanobacteria deliver poisons could possibly colonize generation frameworks, specifically, outside tanks.

Human poisons delivered by cyanobacteria are found in freshwater creatures, ocean water, and estuaries and incorporate epatotoxins, cytotoxins and neurotoxins dermatotoxinas among others. Ecotoxicity of algal poisons seen in herbivorous fish, spineless creatures, for example, molluscs or Daphnia and shrimp; Toxins can influence the imperativeness, development, and richness of numerous life forms. The synthetic structures of crisp water poisons are most likely more assorted than those of marine poisons, between alkaloids, phosphate esters, macrolides, diaryl-lactones and pentachlorides, and epta-peptides. They have been reported types of algae that create poisons in euglenophyceae, Bacillariophyceae, Dinophyceae, Haptophyceae, and raphidophyte divisions. Some cyanobacteria are likewise the generation of poisons and are presumably in charge of the creation of a large portion of the poisons of freshwater algae caused by the multiplication of unsafe algae. The harmfulness of a few mixes may surpass the remedial. Allergens and aggravations are likewise created by some algae. The creation of poisons in some cyanobacteria is affected by natural factors and rivalry, in spite of the fact that the physiological and environmental generation of poison causes are to a great extent obscure. The blossoming of unsafe algae that deliver poisons are not by any means the only wellspring of algae poisons or algal poisons are constantly connected with sprouts. What's more, blossoms can't be precisely anticipated in indigenous habitats, and this additionally applies to the opening of biofuel edit frameworks.

The two types of freshwater and marine algae create poisons can colonize generation frameworks. The present condition of information about the structure of the phytoplankton group is not adequate to foresee whether the poison delivering strains may attack and flourish in algae biofuel generation frameworks regardless of the possibility that such frameworks are at first or ceaselessly sown with nontoxigenic algae strains.

Mixes as of now not known to be unsafe to their essence at low fixations in the generation of low-force algae and low-power biomass may have unfriendly impacts when concentrated 100,000 times amid gathering and drying stages. Concentrated culture techniques can prompt the distinguishing proof of already obscure poisonous material. On the off chance that algae lipids are utilized as a part of significant worth included coproducts, the nature of these items ought to be checked.

Open outside creation frameworks are probably going to build up a few algae populaces. Control of algal arrangement is basic to keeping up the coveted qualities for the treatment of biomass energizes, co-items by guaranteeing that lipid algae removes are alright for utilize, limiting downstream impacts of water-dissolvable poisons.

Human and Animal Pathogens in Algal Cultivation Systems

Spirulina Colta, Chlorella, and Haematococcus were utilized for sustenance creation, and the nearness of pathogens was not of concern. In any case, algae creation frameworks are different groups that can contain pathogenic specialists, particularly if urban wastewater, wastewater from concentrated creature encourage operations (CAFOs), biosolids (cleansing), and excrement are utilized as wellsprings of water or supplements. In spite of the fact that algae development frameworks utilizing wastewater are like a great many seawater purging lakes in the United States, there might be a few word related exposures in light of the fact that algal biomass that is overseen in the generation of ocean growth biofuels is more prominent (i.e., energizes in algae reflux) and Greater focus (first burning and drying for fills). The thickness and likelihood of specific pathogens in wastewater are corresponded with the level of treatment with a more noteworthy number of pathogenic specialists and assorted variety chiefly of wastewater treated in auxiliary treatment. Essential treatment is strong affidavit, auxiliary treatment is the evacuation of natural issue in suspension and arrangement, and tertiary treatment is the expulsion of inorganic segments, for example, nitrogen and phosphorus. Biosolids (sewage slime) it is normal that the first waste water containing in any event these species. In another investigation of treated and biosolid wastewater in Michigan, adenovirus, enterovirus, and norovirus at 100, 70 and 10 percent of the specimens were recognized separately. Ordered personality and wealth of pathogens in biosolids (and, by expansion, release waters) are controlled by the occurrence of contamination in the group to create wastewater and waste water treatment process specifically. We have examined the survival of some biosolid pathogens in the dirt, however the survival of human and creature pathogens in biofuel algae crops is starting to be researched.

At the point when pathogenic operators are available in algae societies, they may have impacts on wellbeing at work or ecological impacts (if discharge happens). The nearness of coliform microscopic organisms or other pathogenic specialists would confine the choices for co-items.

Opportunity mitigation

The choice of known strains that deliver poisons does not alleviate poisons in shut frameworks and moderate poisons worries for open frameworks, despite the fact that poisons might be: Minimizing wellsprings of human and creature pathogens to develop algae may include:

- An abnormal state of cleansing or wastewater treatment. For instance, Phycal squander water goes through a 0.2-micron channel before utilize and utilizes bright light for beginning treatment and reused water treatment. Reuse of waste water for the creation of algae biofuels could take after built up gauges for squander water reuse.
- Use rural manure review (e.g. Sapphire Energy, Inc.).
- Using high pH solvents to decrease the pathogenic survival of cyanobacterial contenders (e.g. Phyco Biosciences)

MOSQUITO BORNE DISEASE

Potential Environmental, Health, and Social Acceptability Effects

The consequences for wellbeing and social worthiness of algal biofuels could be influenced if open lakes are ineffectively overseen and give environment to mosquito hatchlings. Photoreactors and inclines don't speak to the mosquito environment unless it happens and considerable liquid spillage development puddles. Mosquitoes lay their eggs in dormant water shrewdly, which can extend from huge lakes to little lakes or basins. The aggregate surface zone of the algae lake culture is, not the ideal environment in view of the vital fomentation and unsettling for the correct blend of supplements and introduction to light. The females of most mosquito species exhaust the eggs seldom surging or fomented water. Furthermore, moving water may meddle with the surface pressure required for mosquito siphon respiratory capacity. Be that as it may, all the generally still open lakes (bank-like streams and long-running ebbs and flows streaming) and the free-water pools or repositories would be appropriate for the development of mosquito hatchlings. Since algae are sustenance for mosquito hatchlings, the supplement frameworks and high algae supplement culture (when and where water is moderately fixed) might be the primary territory. The turbidity of some cultivating frameworks gives asylum to visual predators. Mosquito species can endure less salt than freshwater conditions, however a few species can endure saline 100 percent ocean water.

Giving natural surroundings to mosquitoes could be a worry for human wellbeing and the adequacy of the creation of algal biofuels for a few reasons, including:

Mosquitoes are viewed as a parasite and a disturbance that can't go on without serious consequences by individuals living close plant development. Districts close wet zones proposed locales at times fabricated restrict in view of the expectation of a mosquito position issue.

Mosquitoes are transporters of numerous human irresistible ailments in the United States, for example, Equine Eastern Encephalitis, Encephalitis Crosse, St. Louis encephalitis, Western Nile Virus, Western Equine Encephalitis, and Dengue Fever. Western Nile infection is additionally anticipated that would be a factor in the decrease of natural life.

If algae lakes wind up plainly rich ground for mosquitoes, there is a hazard that hatchlings turn into a parasite, lessening the algae populace thickness underneath monetarily profitable levels through predation.

Indicators of Sustainability

Supportability markers of mosquito-encouraged infections are the thickness of mosquito hatchlings in lakes and changes in the frequency of mosquitoes owing to ailments.

CHAPTER 5

ALGAE OILS AS FUELS

Algal oil is a biofuel that exists in more than one path to its green charging. It is an emerald-green raw petroleum and originat by the photosynthesis in green growth. This biofuel may fuel autos, trucks, and flying machine without depleting our sustenance delivering crops. "This item can run directly into the present oil pipeline," San Diego claims Jason Pyle of Sapphire Energy in, California. They say that "green rough" is comparable in quality to normally happening raw petroleum. It is started as a side-effect of photosynthesis by a hereditarily built strain of green growth presented to daylight and are housed in the tanks of treated waste-water. The tanks can be set on non-arable land. Stream fuel and diesel gas and have just been refined from the green rough, and the organization is meaning to create 10,000 barrels of biofuel every day inside five years.

BIOCHEMISTRY TOWARD LIPID SYNTHESIS

Green growth are adiverse gathering of life forms that occupy an extensive scope of environment, from the to a great degree frosty (Antarctic) to a great degree hot (leave) locales of the earth. Green growth represent the greater part the essential efficiency at the base of the natural way of life. Lipid digestion (the biosynthetic pathways of unsaturated fats and triacylglycerol, or TAG blend), especially in green growth, has been less concentrated than in higher plants. In view of the succession homology and some mutual biochemical attributes of various qualities and additionally catalysts disconnected from green growth and higher plants that are detailed in lipid digestion, it is by and large trusted that the fundamental pathways of unsaturated fat and TAG biosynthesis is green growth are specifically closely resembling higher plants. The once more blend of unsaturated fats in green growth happens basically in the stromal and thylakoid area of the chloroplast. Green growth settle CO2 amid the day through photophosphorylation (thylakoid) and formed sugar amid the Calvin cycle (Stroma), which changes over into different items, including TAGs, contingent upon the types of green growth or particular conditions relating to plastid and cytoplasm. The biosynthetic pathway of lipid in green growth happens by four stages: blend of palmitic corrosive lastly sugars aggregation inside the cell, amalgamation of higher unsaturated fat by chain stretching, the formation of acetyl-CoA took after by malony-CoA.

Formation of Acetyl-CoA/Malonyl-CoA

Photosynthates give an endogenous wellspring of acetyl-CoA by enacted acetyl-CoA synthetase in the stroma, from the cytosolic change of glucose to pyruvate free acetic acid derivation amid glucolysis. This acetyl-CoA is specially transported from the cytosol to the plastid, where it is changed to the unsaturated fat and along these lines to TAG, which again is transported to the cytosol and structures the lipid bodies. This acetyl-CoA pool will be kept up by the Calvin cycle, pyruvate kinase (PK) and glycolysis interceded union of pyruvate from PEP, which happens in the chloroplast notwithstanding the cytosol. The main response of the unsaturated fat biosynthetic pathway towards the development of malonayl-CoA from acetyl-CoA and CO2 is catalyzed by the chemical Acetyl-CoA carboxylase (ACCase). Figure 5.1, to illuminate the change of acetyl-CoA to malonyl-CoA through using ATP. Amid this procedure, seven atoms of CO2form seven particle of malonyl-CoA and seven particles of acetyl-CoA. This malonyl-CoA experiences amalgamation of long carbon-chain unsaturated fats by rehashing multi-step arrangement, as appeared in figure 5.1 and 5.2. A soaked acyl gather created through this arrangement of responding turns into the substrate buildup by an enacted malonyl gathering.

$$7\ \text{Acetyl CoA} + 7\ CO_2 + 7\ \text{ATP} \quad 7\ \text{malonyl CoA} + 7\ \text{ADP} + 7\ P_i + 14\ H^+$$

Seven cycles of Condensation and reduction

$$\text{Acetyl CoA} + 7\ \text{malonyl Co A} + 14\ \text{NADPH} + 20\ H^+ \rightarrow \text{palmitate} + 7\ CO_2 + 14\ \text{NADP}^+ + 8\ \text{CoA} + 6\ H_2O$$

The overall stoichiometry for the synthesis of palmitate is

$$8\ \text{Acetyl CoA} + 7\ \text{ATP} + 14\ \text{NADPH} + 6H^+ \rightarrow \rightarrow \text{palmitate} + 14\ \text{NADP}^+ + 8\ \text{CoA} + 6\ H_2O + 7\ \text{ADP} + 7\ P_i$$

Figure 5.1: Formation of Seven Malonyl-CoA molecules

Synthesis of Palmitic Acid:

After the development of seven malonyl-CoA particles, a four-stage rehashing cycle (augmentation by two carbon/cycle), i.e., lessening, buildup, lack of hydration and decrease, happen for seven cycles and structures the foremost result of the unsaturated fat synthase frameworks, i.e., palmitic corrosive, which is the antecedent of other long-chain unsaturated fats mixotrophic and heterotrophic are accounted for to create algal fuel within the sight of light. Likewise, the dim hetrotrophic sustenance system is additionally observed to be equipped for lipid biosynthesis through microalgae under the specific conditions.

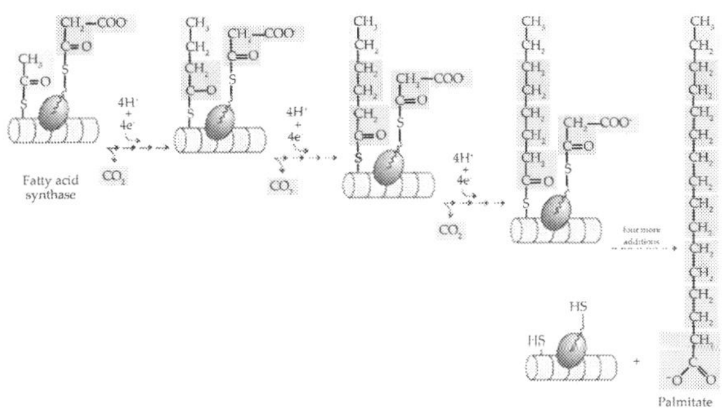

Figure 5.2: Overall process of Palmitic Acid

Source: http://reasonandscience.heavenforum.org/t2168-the-amazing-fatty-acid-synthase-nano-factories-and-origin-of-life-scenarios

Photoautotrophic Mechanism

The basic methodology for the cultivation of microalgae is autotrophic mode. Microalgae in photoautotrophic sustenance mode utilize daylight as the vitality source to shape biochemical vitality by photosynthesis. This is one of the predominant natural conditions for the continuous development of microalgae. In the photoautotrophic nourishing mode, photosynthetically settled CO2 as glucose fills in as a sole vitality hotspot for every metabolic movement. The less complex type of photosynthate, for example, less difficult sugar, fills in as sole vitality hotspot for completing the metabolic exercises of the algal cells. These starches, stretch conditions, and under nourishment restricting will support the lipid biosynthesis, which likewise adapts up to the anxiety. Lipid profitability for the most part relies upon the photosynthetic action as far as climatic CO2 obsession and microalgae species. Extensive process in lipid efficiency, going from 5% to 68% was accounted for under change working conditions species assorted variety. A noteworthy favorable position of the autotrophic dietary mode is the algal oil creation at the use of air CO2. Expansive scale microalgae development framework, (for example, open/raceway lakes) is much of the time worked under photoautotrophic condition. The autotrophic nutritious mode additionally has some defilement issues contrasted and another phone amid dull heterotrophic operation. Glucose transport framework in the algal cell wind up plainly wasteful within the sight of light, due to higher accessibility of photosynthates inside the cell are adequate to photosynthesis and down-direction of hexose transport protein. The carbon is acquired from outside the cell and changed over to the acetyl-CoA pyruvate, which additionally changes over to malonyl-CoA and along these lines, enters the lipid biosynthetic pathway. In heterotrophic nourishment mode, as a result of other glucose accessibility, other metabolic procedures and breath don't contend with the lipid biosynthesis, not at all like autotrophic mode. Moreovermicroalgae can use natural carbon under dull conditions due to the capacity of light-autonomous glucose taking up. Henceforth, the lipid profitability is high in heterotrophic nourishment mode.

Heterotrophic is conceivable to have high microalgal biomass thickness equipped for giving a financially mass-scale generation strategy. Photograph Heterotrophic sustenance mode dodges confinements of light reliance, which is the principle obstacle to get high cell thickness in substantial scale bioreactors. Protothecoides Chlorella indicated higher lipid content (40%) amid heterotrophic development. Also, expanded lipid efficiency (3.700 mg/l/d) was accounted for utilizing a superior sustenance well-disposed clump mode sustaining technique, where lipid profitability was 20 times that got in photograph carrotrophic culture. The primary preferred standpoint of heterotrophic sustenance mode encourages decontamination together with lipid efficiency, giving favorable position over its application to the current situation with the expanded contamination stack. What's more, benefit, its simplicity of operation and simple support are the principle fascination of the Heterotrophic development approach. Be that as it may, heterotrophic frameworks experience the ill effects of contamination issues.

Mixotrophic Mechanism

Microalgae can likewise work in the mixotrophic sustenance demonstrate consolidating both autotrophic instruments than heterotrophs. It encourages the obsession of air CO2 and expends the natural particles of the earth and developing micronutrients. The microalgae can acclimatize the accessible natural mixes and climatic CO2 as acarbon source in the mixotrophic method of sustenance. Other than the method of photograph nutritious heterotrophs is utilize carbon as CO2. Mixotrophs can utilize natural carbon; therefore, the vitality of light is not a restricting element for the development of biomass. The acetyl-CoA pool will be kept up by both carbon sources than by methods for CO2 obsession (Calvin cycle) and presumption from outside the cell, which can make much more malonyl-CoA. Photosynthetic digestion utilizes light and CO2 for development and natural photosynthesis. In the event that the framework is accessible an outside carbon source, there is lost photosynthetic improvement. Mixotrophic societies indicate decreased photograph restraint and enhanced development rates contrasted with autotrophic heterotrophic and societies. The green growth have the adaptability to change their nutritious mode relying upon accessibility and lighting conditions. On the off chance that basic starches are available in the framework, green growth in autotrophic mode change nourishment for hermotropic vitality investment funds, at a slant scenedesmus effectively adjusted to heterotrophic development on glucose in dull conditions. Heterotrophic cells contrast altogether from photoautotrophic cells to different physiological, for example, the fotosensimilación CO2 recurrence and the rate of joining of carbon focus and chlorophyll property. Algal cells are a typical sharing phototrophic heterotrophic oxidation cell pool and connected with mixotrophic method of operation. Microscopic organisms appear to assume an auxiliary part in the natural lessening of oxygen in high-recurrence oxidation lakes, and its part is most likely constrained to the debasement of biopolymers, along these lines delivering the substrates for the utilization of green growth.

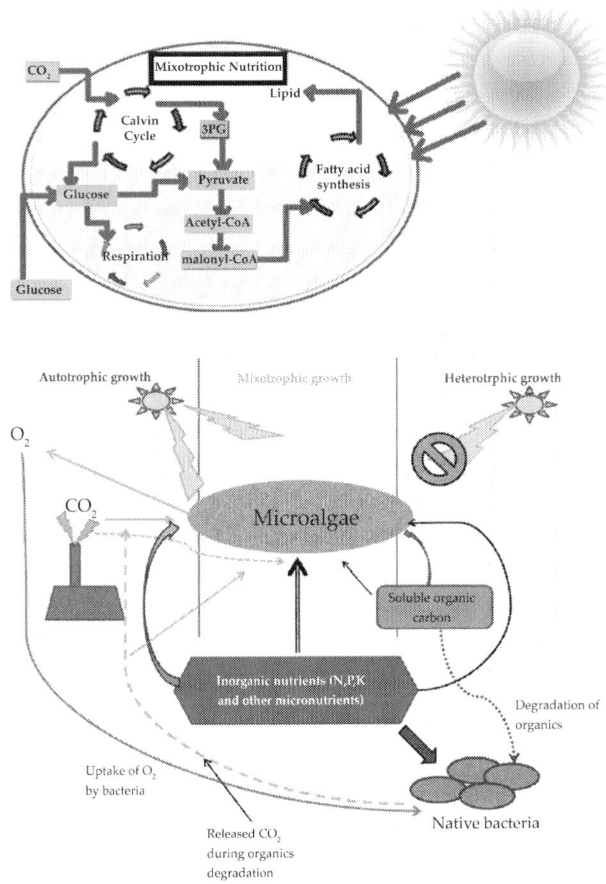

Figure 5.3: Mixotrophic Mode of Nutrition in algal cells toward CO_2 Fixation and glucose assimilation for lipid biosynthesis.

SOURCE: https://www.e-algae.org/journal/view.php?doi=10.4490/algae.2016.31.8.28

The advantages of mixotrophic nourishment are its autonomy as far as photosynthesis as development substrates. The mixotrophic development regimen is a variation of the heterotrophic development administration where CO2 and natural carbon are all the while treated and elements of respiratory and photosynthetic digestion all the while. The mixotrophy regularly saw in water bodies' environmental action where the structure and capacity of living homeostasis frameworks are upheld by science, physical and hierarchical biota that equilibrates the biological state.

The advantages of mixotrophic nourishment are its freedom as far as photosynthesis as development substrates. The mixotrophic development regimen is a variation of the heterotrophic development administration where CO2 and natural carbon are at the same time treated and elements of respiratory and photosynthetic digestion at the same time. The mixotrophy frequently saw in water bodies' biological movement where the structure and capacity of living homeostasis frameworks are bolstered by science, physical and authoritative biota that equilibrates the natural state. The water biological system is for the most part comprised of supplements and natural carbon as indispensable parts, where microalgae, alongside other living segments, cooperate in beneficial interaction. A few types of microalgae are not by any stretch of the imagination mixotrophic however can switch amongst phototrophic and heterotrophic metabolites, contingent upon the ecological conditions. Microalgae gather lipids, are for the most part a development in normal waters; Therefore, natural water bodies incorporated with various types of microalgae can be considered as a potential store for the utilization of biodiesel. In such manner, we endeavored to investigate the limit and capability of blended societies of microalgae got from various streams in the extraction of lipids, which can be further transesterified biodiesel. The investigation additionally centered around how to deliver modest lipid refinement. It has been demonstrated that green growth development was more prominent in mixotrophic conditions, more noteworthy biomass profitability under photoautotrophic conditions. It has been demonstrated that mixotrophic development was a system for high biomass and high development rates, with the additional advantage of delivering photosynthetic metabolites. Solazyme an inexhaustible oil organization In the United States, it built up a dim kelp culture coordinated with heterotrophic process systems, giving carbon sources remotely. The organization is utilizing different types of waste materials as crude material for the development of green growth in fermenters and the utilization of oil up to 75% in light of the dry weight. The organization is reckoning the offer of kelp oil to business refineries until the end.

SUBSTRATES FOR MICROALGAE GROWTH AND LIPID PRODUCTION

CO2

Settling CO2 from microalga through photoautotrophs instruments to draw fluid fuel is viewed as a solid and supportable way to deal with carbon balance. Microalgae are viewed as more photosynthetically effective than earthbound plants to settle CO2. Microalgae likewise have the useful limit of settling barometrical CO2 and modern discharges. In the settling procedure, utilizing CO2 as an inorganic carbon source, while water goes about as an electron benefactor for nourishment stockpiling, holds sugar, lipids that wind up noticeably under certain anxiety conditions. Numerous types of microalgae can utilize sugars, for example, Na2CO3 and NaHCO3 for cell development. Most green growth and cyanobacteria have diverse CO2 focus instruments and go about as inexhaustible fills with CO2 obsession in a brought together and feasible approach. Be that as it may, algal cells can't effectively trap barometrical CO2 to help the fast development required for business operations.

Microalgae can develop in a domain rich in supplements and aggregate supplements and metals from wastewater. This makes heterotrophic microalgae culture a decent alternative for lipid biosynthesis. The generation of ocean growth biodiesel is considered financially and ecologically reasonable when water as a substrate utilized. The utilization of green growth as a biocatalyst is by and large archived for filtration in customary oxidation lakes, repositories, and suspended green growth tubes to expel high groupings of supplements, particularly to polish purposes. The green growth based treatment framework is successful in expelling supplements from wastewater contrasted and substance treatment and is naturally satisfactory and gives reusing supplement proficiency. They are seen most importantly in oxidation lakes. Especially for the treatment of mechanical waste water recovery process green growth preparing database has been utilized as a working tertiary unit for the expulsion of substantial metals and natural poisons rather than supplements. The development of microalgae wastewater treatment is a potential for ecological supportability and the carbon lack of bias alternative.

Five particularly extraordinary natural bodies (Mixotrophic) were assessed to assess the biodiesel creation limit of their local microalgae (blended). The lipid yield shifts from 4 to 26%, depending for the most part on the nature and capacity of the sensibly fuel-productive fuel washer, fuel-proficient wood body with higher body soaks. The green growth profile of assorted variety was the nearness of high lipid gathering species. The prevalence of mixotrophic microalgae (Chlorella, Scenedesmus, and Euglena) and heterotrophic facilitators (focal diatoms and pennatees) was seen, alongside the few photoautotrophs (Spirogyra). Euglena can go about as a carrotroos (day) and a heterotrophic (night). Scenedesmus sp. is by and large engaged with the procedure of normal refinement and dominance in all societies is a positive sign for sewage treatment. Chlorella, Euglena, and diatomas are likewise known to be able to utilize natural carbon show in wastewater with climatic CO2. Biological water bodies can be considered as possibilities for bioenergy creation, joining with the regular purging procedure.

Processed profluent cover reported as a potential hotspot for the generation of algal biomass related with the creation of biodiesel. Two agro-modern cuts, anhydrous dry ethanol, and whey of soybean drain, as feedstock for the development of mixotrophic/heterotrophic microalgae for creation development fuel, is also regarded as waste regarded as a tertiary treatment. The useful part of full scale/micronutrients in carbon, nitrogen, phosphorus, and potassium in the heterotrophic (blended) microalgae culture in purging on the development of biomass efficiency and lipid stages utilizing successive development and craving have been considered. Restricting supplements amid starvation demonstrated a beneficial outcome on lipid efficiency. The states of hardship of supplements caused a diminishing in the substance of the tilacoidal cell film by actuating acyl hydrolysis and invigorating the hydrolysis of the phospholipids. These progressions increment the intracellular substance of CoA. TCA acyl corrosive unsaturated fats to invigorate glucose blend. Label amassing in light of ecological anxiety presumably happens as a methods for giving a vitality tank that can without much of a stretch be catabolized because of a more ideal condition to permit quick development. Joining of microalgae culture with biohydrogen creation demonstrated lipid efficiency for biodiesel generation alongside encourage treatment.

MICROALGAE CULTIVATION

Autotrophic microalgae are developed in expansive bowls or in shut photobioreactors, utilizing CO2 improved. CO2 can come as smoke from control plants or can be gotten from different energizes and petroleum derivatives organic procedures. Along these lines, they can help reuse this particular ozone depleting substance and can help lessen ozone depleting substance emanations by and large when algal biomass moves toward becoming biofuels. Heterotrophic microalgae are developed in substantial fermenters with sugar or starch, for example, corn liquor aging and gives almost 10 percent of our fluid fuel transportation. Green growth (microalgae) are developed in ocean water, as a rule in nearshore frameworks, albeit developing in the vast ocean has been examined in the past are still of intrigue, and even green growth cultivating is a possibility. The green growth can play rapidly, quicker than some other plant, and there are a huge number of types of green growth, with all the more reliably found.

Algae for biofuels

Microalgae are a huge and differing gathering of amphibian life forms and don't have the complex cell structures and are available on the upper planar gathering. They can be found in various situations, a few species prosper in new water while the others are the occupant of salt and ocean water. Most species are photoautotrophic and change over sun oriented vitality into substance shapes through photosynthesis.

There are two fundamental other options to the development of photoautotrophic green growth driving and photobioreactor tanning frameworks (PBR). Since the run of the mill lake drive incorporates a shut oval channel of ~ 0.25 to 0.4 m profound, outside and blended with a water course impeller and counteracts sedimentation (lakes stay retained and shallow shading cells green growth, as optics restricts the entrance of light through the ocean growth stock). In PBRs the way of life medium is encased in a straightforward network of tubes or plates and microalgae juices is flowed from a focal document. PBR frameworks permit better control of the green growth culture condition yet have a tendency to be more costly than conductive lakes. Assistant power request may likewise be higher. The apparent capability of microalgae as a wellspring of ecologically reasonable transport fuel is a solid advancement motor and gives the basic method of reasoning to a significant part of general society bolster for miniaturized scale green growth R and D. It is along these lines critical that biofuel frameworks green growth can unmistakably show their long haul monetary and natural qualifications. Here we take a gander at three parts of microalgal generation eventually, decide future financial suitability and natural supportability:

Vitality adjust and carbon adjust, natural effect and creation costs. Look at each of these perspectives, thus, gives the structure to this report. The investigative approach embraced joins methodical survey and meta-examination with the learning obtained by the workshop of specialists accumulated in 2010 and 2011 as a major aspect of a FP7 European research extend: Aquafuel.

The vitality and carbon adjust of microalgae generation

In the event that microalgae are an imperative crude material for the generation of biofuels, the aggregate vitality (and carbon adjust) must be good. There have been many endeavors to assess this for huge scale generation of microalgae biofuels utilizing LCAs to depict and evaluate data sources and outflows in the creation procedure. Notwithstanding, endeavors have been hampered by the way that there is still no mechanical scale process particularly intended for the generation of biofuels. Subsequently, ACV microalgal bolster information is separated from research facility or business scale frameworks regimens that are intended to deliver high-esteem items, for example, shades and wellbeing sustenance supplements. Regardless of this confinement, it is normal that LCA can even now fill in as a device to help with framework outline.

Table 5 (A): Different studies to assess the life cycle of the algae derived fuels.

Lead author	Description
Kadam	Compares a traditional coal-fired power station with one in which coal is co-fired with algae cultivated using recycled flue gas as a source of CO_2. The system is located in the southern USA, where there is a high incidence of solar radiation
Jorquera	Compares the energetic balance of oil rich micro algae production. Three systems are described: raceway ponds, tubular horizontal PBR, and flat-plate PBRs. No specific location was assumed. The study only considers the cultivation stage and the system energy balance
Campbell	Examines the environmenral impacts of growing algae in raceway ponds using seawater. Lipids are extracted using hexane and then transesterified. The study is located in Australia, which has a high solar incidence, but limited fresh water supply
Sander	A well-to-pump study that aimed to determine the overall sustainability of algae biodiesel and identify energy and emission bottlenecks. The primary water source was treated wastewater and was assumed to contain all the necessary nutrients except for carbon dioxide. Filtration and centrifugation were compared for harvesting. Lipids were extracted using hexane, and then transesterified
Stephenson	A well-to-pump analysis, including a sensitivity analysis on various operating parameters. Two systems were considered, a raceway pond and an air-lift tubular PBR. The location of the study is in the UK, which has lower solar radiation than the other studies
Lardon	Considers a hypothetical system consisting of an open pond raceway covering 100 ha, and cultivating Chlorella vulgaris. Twp operating regimes are considered: i)normal levels of nitrogen fertilization; ii) low nitrogen fertilization. The stated objective was to identify obstacles and limitations requiring further research
Clarens	Compares algae cultivation with corn, switch grass and canola (rape seed). The study was located in Virginia, Iowa, and California, each of which has different levels of solar radiation and water availability, Five impact categories considered: energy consumption (MJ), water use (m^3), greenhouse gas emissions (kg CO2 equivalent, land use (ha), and eutrophication (kg PO_4)

Creation frameworks were contrasted with respects with the net vitality proportion (NER) of biomass generation. NER here is characterized as the aggregate of vitality utilized for development, collecting, and drying, partitioned by the vitality substance of dry biomass. Each time the NER is lower than the unit, the procedure creates more vitality than it devours. The aftereffects of this correlation are appeared in Figure 5.4. Of the eight waterproof ideas can be seen that you have bring down NER 1. This recommends a positive vitality adjust might be achievable for these frameworks, in spite of the fact that the preferred standpoint is peripheral in the standard case. NER of PBR frameworks is more prominent than 1. The best PBR execution is the level plate framework over the tubular PBR as it profits by an extensive light surface and low oxygen stockpiling. (NER is characterized as the aggregate of vitality utilized for development, reaping, and drying, separated by the vitality substance of dry biomass).

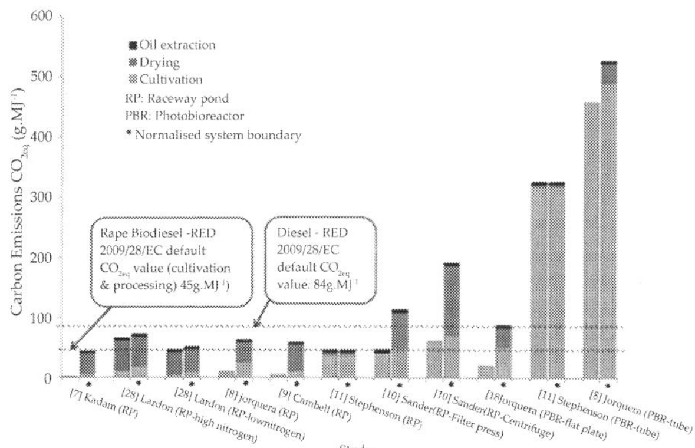

Figure 5.4: Ratios of the net energy required to produce microalgae biomass

The Sander think about utilizations high esteems for the vitality required for development, drying, and gathering in the framework. The investigation exhibited here conveys less vitality yield than that required for contribution to the framework. The first suspicions about the algal species and its efficiency are indistinct so far. The information exhibited above is gotten from the investigations did in the 1980's, and consequently may not be a delegate of later outlines.

The Stephenson contemplate is the main shot that gives an entire portrayal of the development and reaping process, so standardization has any kind of effect for this situation. The vitality necessities of the developing season are higher than different examinations in light of the fact that the creators accept that greater power is required at this phase to conquer grating misfortunes (which it has evaluated from early standards). Less vitality is expected to dry different examinations, concerning consequent preparing steps, creators expect the utilization of an extraction procedure that can acknowledge wet biomass oil (warm recuperation homogenization).

For the pooled tin culture, the best commitment to vitality needs originates from the power required for the dissemination of culture (22% - 79% portion of vitality) and vitality fused into the lake development (part of vitality 8% - 70% of the implicit nitrogen compost vitality can likewise make a generous commitment to the vitality needs (division of vitality for the development period of 6% - 40%) (Note - this range avoids Kadam's examination it incorporates a nitrogen input Mass portion 0.05, an esteem that appears to constantly begin since this investigation shows that biomass contains a mass level of proteins> 30%).All standard PBR frameworks expend more vitality than they deliver.

Drying and drying out of biomass are relatively less essential than vitality utilization for development and reaping. They can achieve higher convergences of biomass in PBR green growth frameworks, and incompletely on the grounds that PBR expends more power amid the way of life stage. The power used to pump culture medium around the PBR and conquer rubbing misfortunes speaks to the majority of the influence ingested amid the way of life stage (the portion of vitality per tubular PBRs is 86% - 92%, the division of PBRs plat Energy is 22%, while the development of the framework speaks to the vast majority of the rest (the part of vitality for the development of the framework is 6% - 12%). Another wellspring of variety is that each examination chooses an alternate piece for the green growth delivered and an alternate efficiency for the development stage; this impacts the vitality required for practical units created. Expecting that green growth efficiency is low, at that point the vitality expected to create 1 MJ of dry biomass will be higher (on the grounds that the blending prerequisite per unit of time won't be diminished). An inconvenience factor is that the development of low profitability green growth, for example, nitrogen, can enable green growth to aggregate more lipids and along these lines can prompt higher calorific power for biomass by and large. It is plainly imperative that the profitability and organization esteems coordinate each other and mirror the working of the framework.

The carbon related with the generation of algal biomass has been evaluated by duplicating the vitality contributions at the procedure with predefined outflow factors portrayed in the European Renewable Energy Directive. The outcomes acquired are appeared in Figure 5.5. It can be seen that most emanations are related with electric power utilization for pumping and blending and giving warmth to dry the green growth. Specifically, discharges related with the generation of algal biomass in lakes are similar periods of harvest emanations and creation of biodiesel rapeseed methyl. In any case, generation in PBRs demonstrates higher discharges than customary fossil diesel. A vital cautioning to this examination is that carbon emanations are very reliant on various vitality utilize discharge factors in the framework (specifically power) and bland elements may not be proper in all circumstances.

(The default discharge factors used to appraise carbon dioxide were: - Diesel 83.80 g MJ-1, power: 91 g MJ-1, warm: 77 g Factor MJ-1 emanation of compost and PVC covering (For the situation of ducted lakes) and PBR is thought to be the same for the warmth).

There was ageneral accord among the specialists being referred to, in any case, that the assessments of the development rate of green growth (both as far as biomass efficiency and lipid yield) aren't right with hopeful esteems and don't consider the misfortunes create a procedure of expansion. Aquafuels Round Table members likewise noticed that biomass efficiency evaluations ought to be founded on yearly normal esteems, bringing up this is not proportional to normal profitability on a late spring day.

PREPARATION OF BIOFUELS

Algal living beings are photosensitive green growth or microalgae that develop in oceanic conditions. Macroalgae or "green growth" are multicellular plants that develop in sweet or salty water. They are arranged into three gatherings in light of pigmentation: dark colored green growth (Phaeophyceae), red green growth (Rhodophyceae) and green growth (Chlorophyceae).

Microalgae are unicellular photosynthetic microorganisms, living in freshwater or saline situations that change over daylight, water, and carbon dioxide into algal biomass. The three most essential lessons as far as theabundance of microalgae are diatomaceous (Bacillariophyceae), green growth (Chlorophyceae) and (Chrysophyceae).

Among the eucaryotic Chlorophyceae microalgae class, the most usually utilized are Chlamydomonas, Chlorella, and Dunaliella and Haematococcus genera. As related sea-going plants, macroalgae flourish in fluid, vaporous societies, where cells approach adequate light, carbon dioxide, and different supplements. Green growth are mostly developed photoautotrophically; yet, a few species can survive heterotrophically debasing natural substances, for example, sugars. Not at all like earthly plants, microalgae don't require fruitful land or water system. Since green growth expend carbon dioxide, substantial scale development can be utilized to cure the ignition debilitate focal terminal. Ocean growth biomass can assume a critical part in tackling the issue of sustenance creation and biofuels sooner rather than later. Microalga seems, by all accounts, to be the main wellspring of biodiesel from inexhaustible sources that can take care of the worldwide demand for transport energizes.

The conceivable outcomes of utilizing green growth oil as a fuel source

Over 80% of the vitality we utilize originates from three non-renewable energy sources: oil, coal and flammable gas. Around 98% of carbon dioxide discharges are produced from non-renewable energy source burning.

Table 5.1: Percentage composition organic dry matter in various species of algae

Species of sample	Proteins	Carbohydrates	Lipids	Nucleic acid
Scenedesmus obliquus	50-56	10-17	12-14	3-6
Scenedesmus quadricauda	47		1.9	
Scenedesmus dimorphus	8-18	21-52	16-40	-
Chlamydomonas rheinhardii	48	17	21	-
Chlorella vulgaris	51-58	12-17	14-22	4-5
Chlorella pyrenoidosa	57	26	2	-
Spirogyra sp.	6-20	33-64	11-21	-
Dunaliella bioculata	49	4	8	-
Dunaliella salina	57	32	6	-
Euglena gracilis	39-61	14-18	14-20	-
Prymnesium parvum	28-45	25-33	22-38	1-2
Tetraselmis maculata	52	15	3	-
Porphyridium cruentum	28-39	40-57	9-14	-
Spirulina platensis	46-39	8-14	4-9	2-5
Spirulina maxima	60-71	13-16	6-7	3-4.5
Synechoccus sp.	63	15	11	
Anabaena cylindrica	43-56	25-30	4-7	-

Less measures of petroleum derivatives are utilized to diminish the measure of carbon dioxide and different poisons produced. This can be accomplished by utilizing less vitality or non-renewable energy sources by supplanting sustainable fills. Sustainable power source is a promising option arrangement that sets CO2 in the environment through photosynthesis.

Table 5.2: Oil contents of some microalgae strains

Microalgae	Oil content (wt% of dry basis)
Botryococcus braunii	25-75
Chlorella sp.	28-32
Crypthecodinium cohnii	20
Cylindrotheca sp.	16-37
Dunaliella primolecta	23
Isochrysis sp.	25-33
Monallanthus salina	>20
Nannochloris sp.	20-35
Nannochloropsis sp.	31-68
Neochloris oleoabundans	35-54
Nitzschia sp.	45-47
Phaeodactylum tricornutum	20-30
Schizochytrium sp.	50-77
Tetraselmis sueica	15-23

Profitability and yearly oil substance of green growth are substantially higher than seed crops. Soy may just create around 450 liters of oil for each hectare. Canola can create 1200 liters for each hectare, and the palm can deliver 6000 l. Green growth, then again, can create 90000 liters for each hectare. Microalgae contain lipids and unsaturated fats as parts of the layer, stockpiling items, metabolites, and vitality sources. Green growth contain in the vicinity of 2% and 40% lipid/oil weight (Table 4.1). The green growth can develop anyplace where there is sufficient sun and some may develop in salt water. All green growth contain proteins, starches, lipids and nucleic acids display in various extents. The microalgae can finish a full cycle of development each couple of days. In spite of the fact that rates may fluctuate, there are green growth sorts containing up to 40% of aggregate unsaturated fats. Developing green growth can deliver 30-half oil (Table 4.2). Oil supply depends on signs that could create 47,000 to 308,000 l/ha/year utilizing ocean growth oil.

The most noteworthy yielding green growth

The green growth utilized as a part of the creation of biodiesel are for the most part unicellular oceanic green growth (Chlorophyceae). This sort of green growth is a photosynthetic eukaryote described by high development rates and high populace thickness. In great condition, green growth can twofold their biomass in under 24 hours. Green growth may likewise have high lipid content, ordinarily more than half. This elite is perfect for concentrated cultivating and can be an amazing hotspot for biodiesel creation.

- Chlorella

Chlorella is a unicellular green alga having a place with the Chlorophyceae class. It is an essential variable based math since it develops autotrophically. It is of circular shape, around 2 to 10 microns in distance across, and not flaky. Chlorella has green photosynthetic shades chlorophyll and chlorophyll-b in its chloroplast. Utilizing photosynthesis, it duplicates rapidly by requiring just carbon dioxide, water, light, and a couple of minerals to recreate. It is trusted that Chlorella can fill in as a potential wellspring of sustenance and vitality due to their photosynthetic proficiency to get 8% that is practically identical to other very productive yields, for example, sugar stick. Control of metabolic pathways can divert cell capacity to the blend of favored items and furthermore increment small scale green growth preparing abilities. Heterotrophic microalgae, for outer carbon sources, give a lot of compound vitality, cells frequently put away as lipid beads.

It has been demonstrated that protothecoides Chlorella refined heterotrophically aggregated up to 55% of its dry weight as oil, contrasted and just 14% of the photoautotrophically developed cells.

- Dunaliella

Dunaliella is a green unicellular alga having a place with the class of Chlorophyceae. It's additionally an essential green growth. It is theovalshaped pole of around 9 to 11 microns in width. Life forms are anything but difficult to develop and not assembled or molded chains. The properties of different greasy esters decide the general properties of biodiesel fuel. No exertion or green growth species that can be said to be the best as far as oil execution for biodiesel. In any case, alongside green growth, green diatoms are the most encouraging.

Dimorphus Scenedesmus is a class of unicellular green growth Chlorophyceae (green growth). While this is a most loved animal categories for yielding biodiesel oil, one of the issues with Scenedesmus is that it is substantial, and shape thick silt if not kept in tertiolecta agitación.Dunaliella is a dim green scourge with a size of 10 to 12 micron in width. It is said that this strain has an oil yield of around 37%. D. tertiolecta is a quickly developing strain, enabling you to have a high carbon dioxide rate.

Various processes to convert algae oil into biodiesel

In green growth, light is caught by lightweight complex accumulation proteins, known as LHCI and LHCII. Its demeanor relies upon the ecological conditions (light power). These proteins tie a lot of chlorophyll and carotenoid in the plant and assume a part both in catching light and the dispersal of overabundance vitality which would somehow or another restrain the photosynthetic response focuses (Photosystem II). The excitation vitality used to work the photosynthetic responses is diverted to the photosynthetic response focuses of photosystem I (PSI) and photosystem II (PSII) through the system of the shades, PSII and PSI subunits stuck LHC. To begin with, PSII utilizes this vitality to direct the photosynthetic water part response, changing over water into protons, electrons, and oxygen. Electrons go through the electron transport chain with photosynthetic plastoquinone (PQ), cytochrome b6f (Cyt b6f), PSI and ferredoxin (FD) and NADPH.

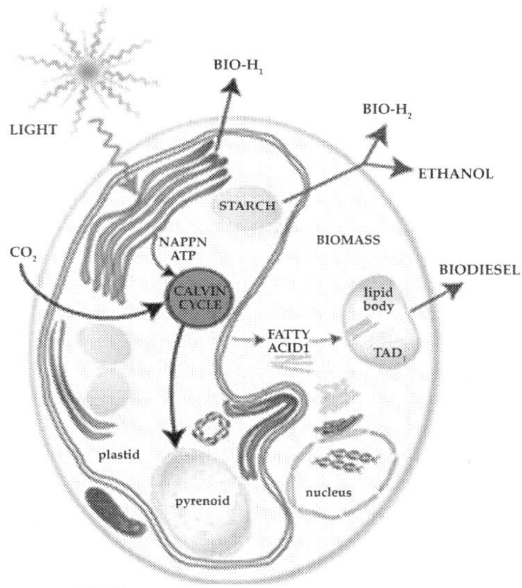

Figure 5.5: A realistic portrayal to represent the outflow of carbon dioxide amid algal biomass generation in raceway lakes. (a)Metabolic pathways included amid the generation of small scale algal bio-fuel. The mix of metabolic pathways is co-ordinated through acomplex instrument that controls photosynthesis yield for the combination of protein, nucleic corrosive.

In the meantime, protons are discharged into the tilioid lumen of PSII and the PQ/PQH2 cycle. This creates a protonic inclination that actuates the generation of ATP by combination. Protons and electrons join from ferredoxin-NADP + oxidoreductase (FNR) to create NADPH. NADPH and ATP are utilized as a part of the Calvin cycle for the generation of sugars, starch, oil required for the creation of bioethanol, biodiesel and biomethane.

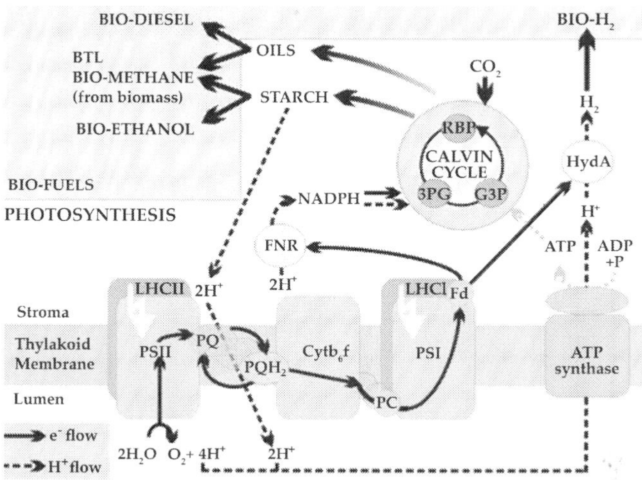

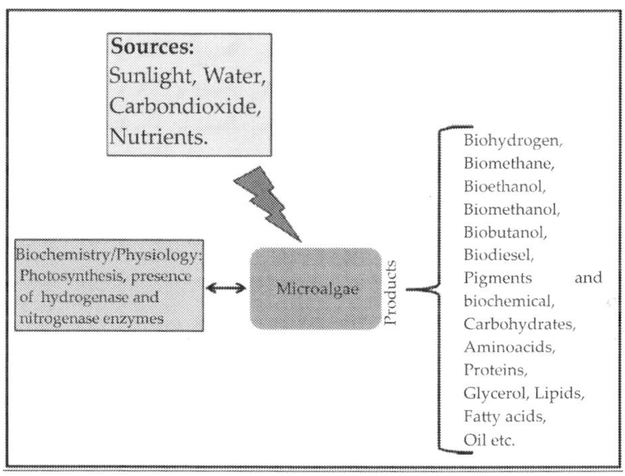

Figure 5.5: (b) Transformation of sun based vitality through photosynthesis into concoction energySOURCE:https://www.researchgate.net/distribution/312188478_Biofuel_production_Challenges_and_opportunities

Green growth oil progresses toward becoming biodiesel through a procedure of transesterification. Oil separated from green growth is blended with liquor and a corrosive or base to create methyl esters of unsaturated fats that make up biodiesel. Overabundance methanol is utilized to constrain the response to support the correct side of the condition. Overabundance methanol is recouped and hence reused. In the event that biomass has developed consistently, its burning does not influence the CO2 adjust in the air, since CO2 radiated by ignition of biomass is balanced by CO2 settled by photosynthesis.

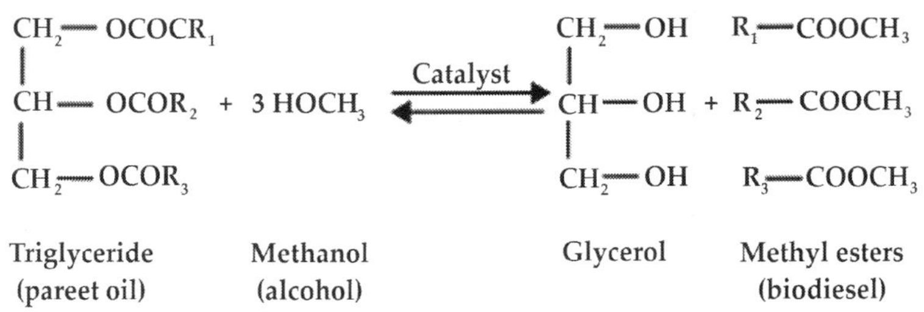

Figure 5.6: Transesterification chemical equations

The generation procedure comprises of microalgae oil creating microalgae biomass which requires light, carbon dioxide, water and inorganic supplements (nitrates, phosphates, and iron) of anhydride. About portion of the dry weight of microalgae biomass is carbon, for the most part got from carbon dioxide. Subsequently, creating around 183 tons of carbon dioxide settled 100 tons of algal biomass. The ideal temperature for becoming numerous microalgae is in the vicinity of 293 and 303 K. A temperature outside this range can execute or harm the cells.

Board/Press, Extraction with Extract and Extraction with Supercritical Fluids are notable to remove oil from green growth techniques. A press/ejector separated 70-75% of green growth oils. The utilization of chemicals, for example, hexane (which are moderately economical) can likewise be utilized to evacuate green growth oils. Supercritical liquid extraction is more effective than the dissolvable expulsion techniques. Since supercritical liquids are specific, and give a high immaculateness item fixation and can extricate just about 100% of the oils. In supercritical liquid extraction (CO2), CO2 is melted under weight and warmed to the point that it has the properties of a fluid and a gas. This condensed liquid at that point goes about as a dissolvable in the extraction of the oil. In the wake of separating the rest of the biomass from the green growth oil, it can be utilized as a high-protein nourishment for steers. This gives more an incentive to the procedure and decreases squander.

Commercial reactors

The greater part of the green growth species are phototrophic and in this manner require light for development. Phototrophic microalgae are generally developed in open tanks and photoreceptors. The untamed water frameworks are shallow lakes in which green growth are additionally developed. Open lakes crops are financially more good yet posture issues of soil cost, accessibility of water and proper climatic conditions. Photovoltaics are distinctive sorts of tanks or shut frameworks where green growth are developed. Photoreactors give a shut social condition, ensured against attacking microorganisms. This innovation is generally costly contrasted with open bowls because of foundation costs. Notwithstanding, shut frameworks require less light and farmland to develop green growth. Species with ahigh substance of microalgae oil become under advanced conditions for development photobioreactor can possibly create 19,000 to 57,000 l of microalgae per section of land every year. A perfect biomass generation framework should utilize the uninhibitedly accessible daylight.

Photobioreactors

The most productive approach to develop microalgae is flowing in huge lakes. Shut sterility photobioreactor and give more prominent control over harvest parameters, for example, light force, carbon dioxide, supplement levels and temperature. Under ideal conditions, microalgae populaces can repeat inside hours and achieve high cell thickness. Notwithstanding sparing water, vitality, and chemicals, shut bioreactors have numerous different advantages, which are dependably the reactor decided for biofuel creation, as their expenses are diminished. Support up to five times more prominent efficiency than the reactor volume and in this manner have an "impression" decreased on the premise of the outcomes. This is ideal on the grounds that the objective is to gather more sun powered vitality in a given field. Most shut photoreactors are outlined as tubular reactors, plates or air pocket reactor segments. To expand effectiveness, photobioreactors ought to be intended to appropriate light finished a wide territory to give direct light power to cells (light weakening). This is normally expert by setting in tubular reactors like a development fence. The boards are arranged north/south to maintain a strategic distance from coordinate light and light strikes the surface. Thusly, daylight is weakened on a level plane and vertically.

A tubular photobioreactor is a progression of straight tubes are generally plastic or glass. This tubular network or sun based authority catches daylight for photosynthesis. The sun powered gatherer's tubes by and large have under 0.1 m in measurement light to enter a huge volume of suspended cells. The ground under the sun oriented authority is painted white or secured with white plastic sheets to expand reflection, in this way expanding the aggregate light got from the channels. For huge scale generation of microalgaebiofuels, there ought to be a developing number of photorepores.

Figure 5.7: **(a)** A tubular photobioreactor **(b)** A closed bioreactor system

SOURCE: https://www.researchgate.net/publication/266211232_Growth_physiology_and_photosynthetic_performance_of_green_microalgae_mass_culture_grown_in_a_thin-layer_cascade

As bioreactors increment in estimate, the level of many-sided quality ought to be decreased to limit the cost per square meter. It is essential to utilize a kind of green growth is both quickly developing amid the progression immunization and exceedingly beneficial in the open tide open entry. This procedure enables low supplement conditions to keep the spread of obtrusive species by empowering the constant generation of green growth biofuels.

Economics of green growth biodiesel creation:

Photoreactors require 10 times more capital speculations than open tin frameworks. The assessed cost for green growth open tank generation frameworks ($ 10/kg) and photobioreactors ($ 30-$ 70/kg) is two higher request biomass amounts separately arrange. Expecting that biomass containing 30% by weight oil and carbon dioxide accessible at no cost, the evaluated cost of creation for photoreactors and lakes would be $ 1.40 and $ 1.81 for every liter of oil individually. Be that as it may, for focused microalgae biodiesel with petrodiesel, the cost of green growth oil ought to be lower than $ 0.48/l. It was evaluated that biodiesel m3 0.53 billion would be expected to supplant the present utilization of transport all energizes oil. It is assessed that the yield per hectare of ocean growth oil is from 18,927 to 75,708 liters for each section of land every year; this is 7-31 times higher than the following best palm oil gather.

To deliver this sum (0.53 trillion) of biodiesel, palm oil would should be developed in a zone of around 111 million (M) hectares. This records for just about 61% of all farmland developed in the United States. Developing palm oil at this scale would be reasonable in light of the fact that insufficient soils stop to deliver sustenance and different yields.

- Algae biomass development has potential:

Biodiesel is as of now created by orchestrated oleaginous regular sun based authority fuel and put away as compound vitality. This speaks to a way for the creation of sustainable carbon and carbon unbiased.

- Advantages of microalgal frameworks:

Microalga Systems:

1) They have agreater effectiveness of photon change (as prove by the expansion in biomass yield per hectare).
2) It can be reaped in groups all year, giving a solid and consistent supply of oil.
3) Salt and wastewater streams can be utilized, accordingly decreasing the utilization of new water.

4) It is conceivable to consolidate generation with CO2 sequestration of unbiased fuel of CO2.
5) For the generation of non-poisonous and very biodegradable biofuels.

The advantages of ocean growth biodiesel:

Oil algae/algae biofuels

1) Accommodates fast development rates.
2) It develops for all intents and purposes all over, a high return for each section of land (7-31 times higher than the following best-gathered palm oil).
3) Biofuel green growth are not lethal.
4) Biofuel green growth are exceedingly biodegradable.
5) High levels of polyunsaturated unsaturated fats in kelp biodiesel are appropriate for frosty atmospheres of time.
6) It is conceivable to lessen carbon discharges in light of where it developed.
7) Seaweed biomass can deliver extensive volumes and this biomass can create a higher level of oil than different sources.
8) Algae oil has little rivalry in the market.
9) Algae can be developed on soils, crisp water, and ocean water.
10) The creation of green growth Innovations will enable it to be more gainful on the assets that would somehow or another be viewed as squandered waste.
11) Have more noteworthy photon change productivity.
12) Can be gathered in groups all year, giving areliable and persistent supply of oil.

- Disadvantages of biodiesel from green growth oil:

Biodiesel from ocean growth oil:

1) Unstable biodiesel creation with numerous polyunsaturated
2) Biodiesel has poor execution contrasted with the principle options.
3) This is a moderately new innovation.

Different restrictions are low groupings of free CO2 required for green growth crest development and herbivorous green growth are a noteworthy however to some degree disregarded issue.

Difficulties in Agriculture:

1) Too coordinate daylight can kill green growth.
2) The temperature must be steady.
3) Crowding hinders the development of green growth.
4) The item "squander" oxygen must be consistently expelled from the water.
5) Open green growth lakes are liable to dissipation and precipitation, causing saltiness and pH awkward nature.
6) Local green growth species develop time after time the coveted pressure.

The troubles in the efficient generation of biodiesel from green growth not by extraction of oil, but rather to locate a high-fat lipid strain and a quick development rate that is not very hard to gather, and a cost photobioreactors culture framework) more appropriate for The strain. It is trusted that the new materials and plans for development in shut bioreactors and the utilization of metabolic designing strategies of craftsmanship and choice/choice are the most encouraging.

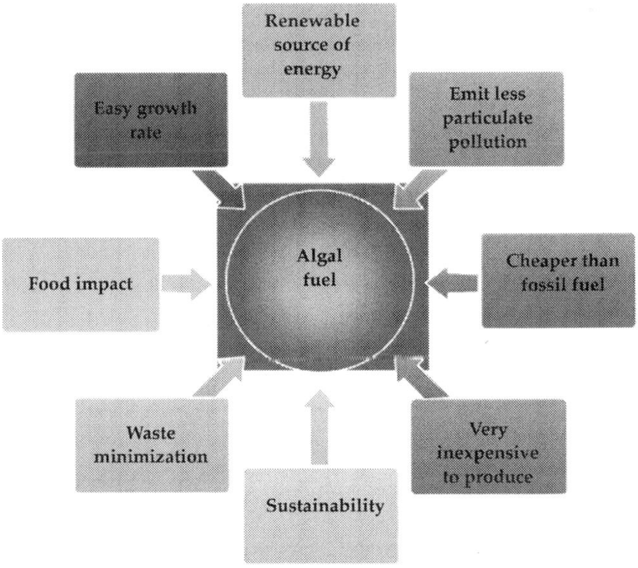

TRANSESTERIFICATION

In natural science, transesterification is the procedure of trade of the liquor gathering of another ester by another liquor. These responses are regularly catalyzed by the expansion of a corrosive or base. The primary response for change of biodiesel oil is called transesterification.

At the point when put in an expansive abundance of a liquor with the nearness of an ester of corrosive or base, there might be a trade of alkoxy gatherings. Over the top liquor mishandle is utilized to drive forward response. The most well-known technique is the transesterification response of the ester with a liquor within the sight of a corrosive impetus

For Example

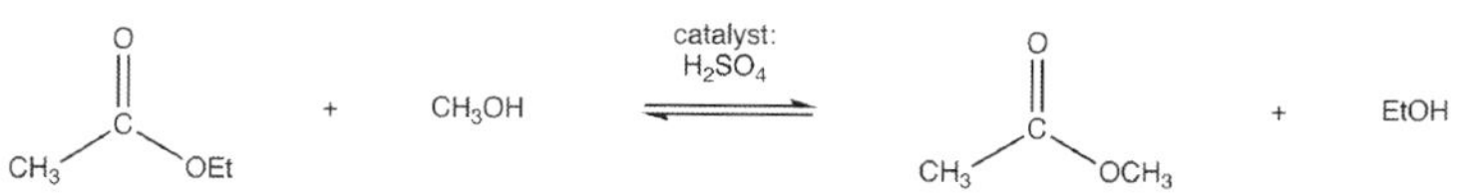

This reaction has the following mechanism:

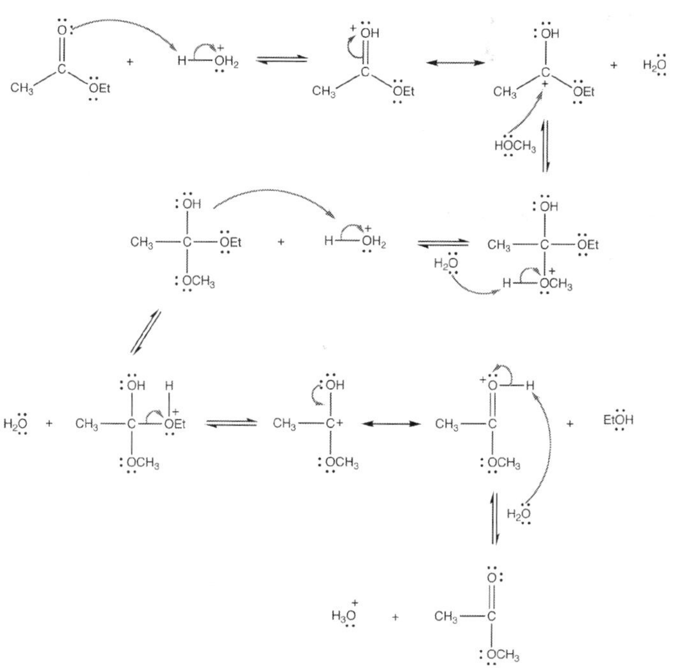

Since the two reagents and items are an ester and a liquor, the response is reversible and the balance steady is near one. Therefore, the rule of Le

Chatelier ought to be misused to convey the response to finish. The least demanding path is to utilize liquor as a dissolvable.

General reaction

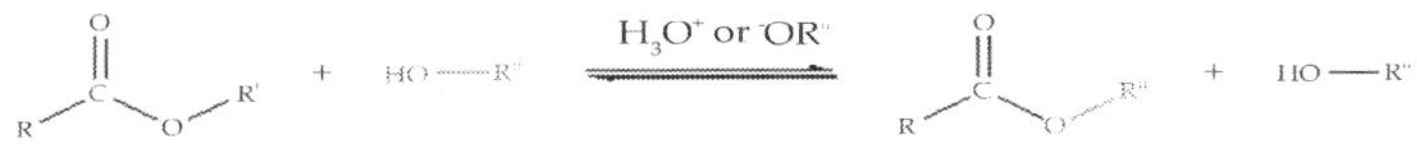

Example 1:

The mechanism in basic conditions

Transesterification system with essential catalysis:

The transesterification response is catalyzed by a base. Any solid base of liquor deprotonation (eg NaOH, KOH, sodium methoxide, and so forth.) yet sodium and potassium are regularly decided for their cost. The nearness of water causes hydrolysis of the undesirable base, so the response must be kept dry.

In the transesterification component, carbonyl ester beginning carbide (RCOOR1) experiences nucleophilic coupling from breathed in alkoxide (R2O-) to give a tetrahedral middle of the road, coming back to the beginning material or continues to the transesterified item (RCOOR2). There are a few animal categories in adjust, and item dispersion relies upon relative reagent and item energies.

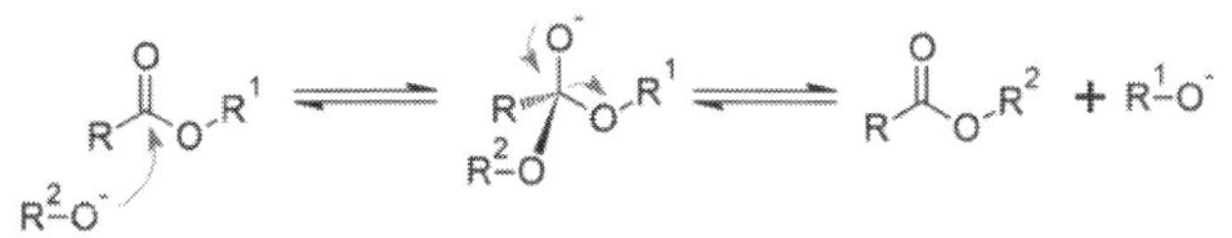

1) **Nucleophilic attack by an alkoxide:**

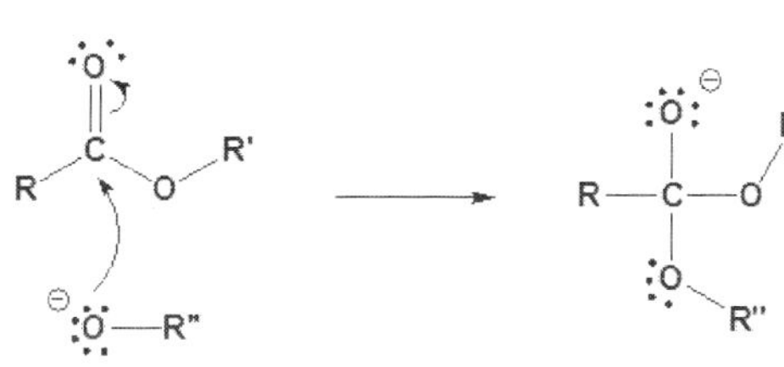

2) Leaving group removal:

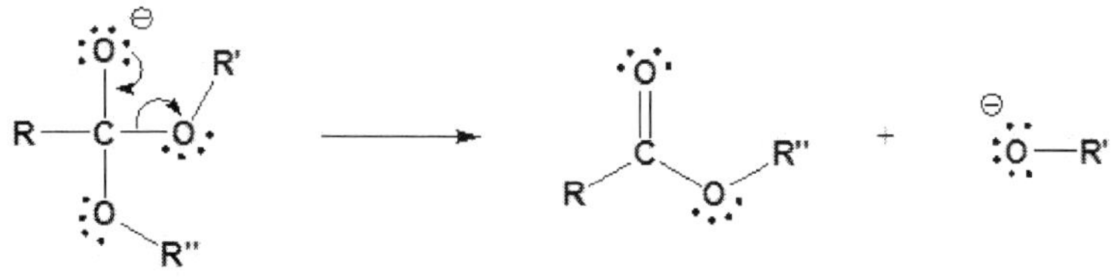

Mechanism in acidic conditions

1) Protonation of the carbonyl by the acid:

The carbonyl is activated towards nucleophilic attack.

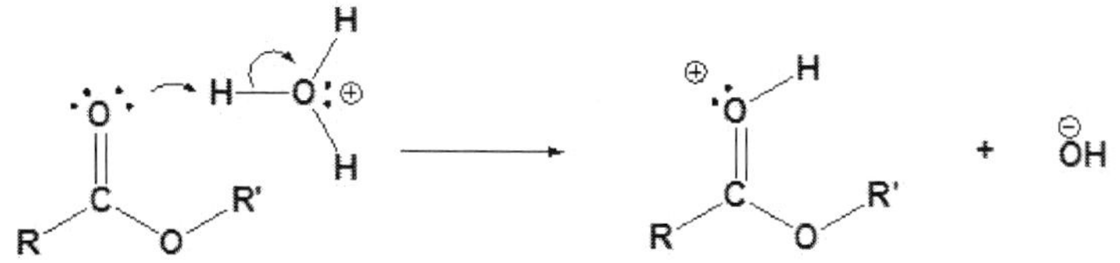

2) Nucleophilic attack on the carbonyl:

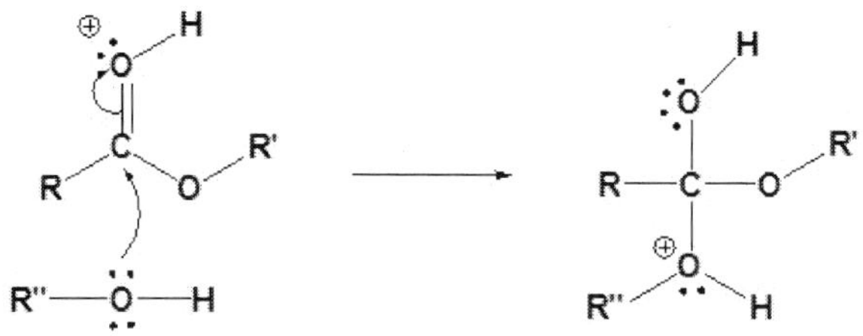

3) Proton transfer:

4) Removal of the leaving group:

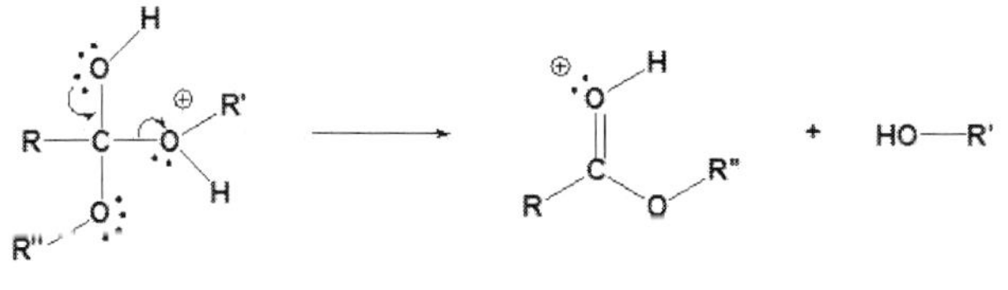

5) Deprotonation:

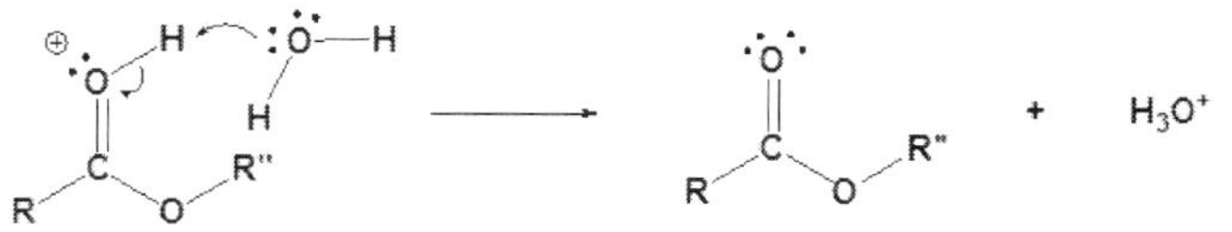

Biodiesel is normally created by the transesterification of vegetable oils, creature fats or kelp crude materials. There are a few techniques for completing this transesterification response including group aggregate process, supercritical procedures, ultrasonic strategies, and even microwave techniques.

Synthetically, biodiesel transesterification containing a blend of mono-alkyl esters of long chain greasy acids.The most normal frame utilizes methanol (changed over sodium methoxide) to deliver methyl esters (regularly called methyl esters of unsaturated fats (FAME)), as it is the least expensive liquor accessible; Although ethanol can be utilized to shape an ethyl ester (ordinarily alluded to as ethyl ester unsaturated fat (FAEE)), higher biodiesel and alcohols, for example, isopropanol and butanol have likewise been utilized. Utilization of higher atomic weight alcohols enhances the subsequent icy coolant properties of the subsequent ester, at the cost of theless effective transesterification response. A procedure of creating transesterification of lipids makes the base oil to the coveted esters. Any free unsaturated fat (FFA) in the base oil progresses toward becoming cleanser or expelled from the procedure, or esterified (creating more biodiesel) utilizing a corrosive impetus. After this treatment, biodiesel has properties fundamentally the same as those of diesel burning and can supplant it in most current employments.

The methanol utilized as a part of most biodiesel generation forms is done with non-renewable energy source inputs. Notwithstanding, there are inexhaustible methanol sources incorporated utilizing carbon dioxide or biomass crude materials, influencing petroleum product to free procedures profitable.

ALGAL FUEL PROPERTIES

The portrayal of green growth oil acquired after transesterification has demonstrated the likelihood as biodiesel. The properties of microalgae oil depend for the most part on the crude material and the transformation strategy utilized. The key perspectives for evaluating the properties of microalgae oil are: the quantity of degrees of corrosiveness, iodine number, particular weight, thickness, kinematic consistency, streak point, sliding point, calorific esteem and octane number.

The physical properties of microalgae oil demonstrate their adequacy for use as biodiesel. Inferred properties of the acridity number (A) demonstrate the destructiveness of the oil; Iodine (IV) alludes to the level of unsaturation. AN and IV recorded inside the cutoff points showed bring down destructiveness and higher fuel immersion green growth. Similarly, a particular gravity and thickness including vitality effectiveness as afuel. The combustibility point communicates the most reduced combustibility point recorded for microalgae oil that decides the oil potential to shape combustible blends with generally bring down ordinary diesel fuel temperatures. The sliding point is the most reduced temperature at which the oil moves toward becoming semi-strong and loses its stream properties. It is additionally a critical parameter of value diesel in tropical nations, for example, India. Diesel oil test microalgae hardening temperature. Like the sliding point, thickness characterizes the stream resistance of the liquids; Calorific power is the vitality discharged as warmth when a compound.

CHAPTER 6

ALGAL BIOMASS HARVESTING

STEADINESS AND SEPARABILITY OF MICROALGAE

The qualities of microalgae and the state in which they create can enormously influence the decision of gathering innovation and execution. Partition of little particles in algae suspension and broke down stock might be bulky in light of the fact that algal cells regularly convey a negative charge and abundance extracellular polymeric substances (EPS) to keep up the soundness of algae in a scattered state. The dependability of microalgae in the development medium is mostly connected with surface algae stacking, the size, and thickness of algal cells, which thus influences the detachability of watery suspensions. Both electrical communications between algae cells as cell associations with the encompassing society juices add to the security of algae suspension, while the size and thickness of algal cells decide their sedimentation rate, which is a critical sedimentation configuration process thought. Collecting expenses might be hoisted on the grounds that mass divisions in the way of life soup are for the most part low.

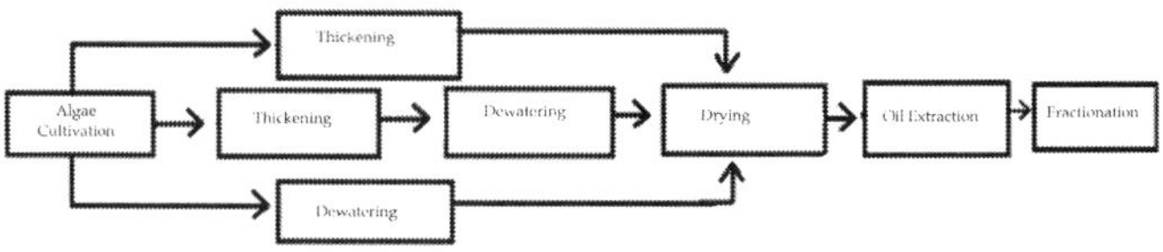

Figure 6.1: Final harvesting products

Concentrates on the impact of the surface heap of particles by different strategies for treatment have been generally reported. Farvardin and Collins (1989) noticed that pre-ozonization builds the surface charge of humic substances. In another examination, resaercher watched an expansion in the soundness of Na-montmorillonite molecule suspension expanded ozone, ascribed to expanded surface charge due to ametal-oxygen breakdown in the crystalline grid. Be that as it may, these investigations have detailed the presence of an ideal ozone dosage through which it can enhance the coagulation of the particles.

Then again, Chheda and Grasso (1994) uncovered that ozonization lessened the dependability of Na-montmorillonite particles covered with normal natural issue (NOM) in waterway waters. It has been proposed that adsorption on NOM Na-montmorillonite particles itself would make the particles more hydrophilic. Be that as it may, resulting ozonization makes the surface of the molecule be contrarily charged less. This, thus, prompted the fractional deallucation of Na-montmorillonite and the change of NOM covered to build the hydrophobicity of the surface of the particles, along these lines destabilizing the particles. It can be reasoned from these investigations that the best possible treatment of ozonization associates destabilizes the particles, prompting an enhanced detachment medium. Systems have been proposed to expand the coagulation of suspended particles caused by ozonization. These components include:

1. Greater carboxylic corrosive substance increments flocculant alumina adsorption and calcium and magnesium accelerate.
2. Molecular lessening of natural adsorbed weight to decrease steric particles.
3. Free of theorgan-metal obligation of free particles, for cxamplc, Fe3+ for natural precipitation.

4. Turbidity and oocyte of algal cells discharge biopolymers for thickening.
5. Polymerization of bigger particles for affidavit.

A specialist watched that rises with changed surfaces with chemicals, both a long hydrophobic tail and a high hydrophilic head stack, can create adequate seabed freedom without flocculation and upstream flocculation. It has been accounted for that the algogenic natural materials (AOM) extricated from these four types of algae (Microcystis aeruginosa and Melosira sp., Chlorella vulgaris, Formosa Asterionella) have been ruled by potential hydrophilic polysaccharides and hydrophobic proteins of low specificity UV absorbance And zeta negative. AOM hydrophobicity is inferable from hydrophobic proteins of more noteworthy than 500 kDa atomic mass. Also, the charge thickness for AOM owing to hydrophilic sugars and humic acids and non-hydrophobic acids diminishes conversely with hydrophobicity. What's more, the restraint of AOM disconnected algae thickening hindrance by Aphanothece halophytic cyanobacilli has been accounted for. AOM is theorized that they can shape complex mixes with ferrum coagulation hence restraining. As examined above, ozonization can expand NOM's hydrophobicity, in this manner expanding coagulation. Be that as it may, the effect of ozonization on OMA is not clear and ought to be additionally examined.

Like planktonic cells, algal cells regularly convey anegative shallow charge. While surface algae charge is gotten from ionic particle bunches in the cell mass of algae and particular adsorption of culture soil particles, theload current is affected by algal species, pH, medium particle focus, and numerous other ecological conditions.

In view of the standards of Deyaguin-Landau-Verwey-Overbeek (DLVO) colloidal dependability hypothesis, communications between colloidal particles are affected by different intuitive powers, for example, twofold layer electrostatic shock, van der Waals fascination, and steric collaboration ¬acción. There is a potential vitality obstruction that should be overcome if wanted to coagulate the moment charge particles. It can be overwhelmed by the motor vitality of the particles, or by diminishing the vitality boundary. This is finished by compacting the twofold layer or expanding the grouping of counter particles or utilizing higher esteem counterins. In spite of the fact that the twofold layer hypothesis is of the hypothetical significance, its utilization is restricted to situations where particular concoction associations accept a section in colloidal reliability.

The destabilization of the colloidal suspension as an item in algae development following particular substance communications can be accomplished by the nearness of natural polymers. Business polymers, for the most part those of high polyelectrolyte sub-atomic weight, for example, polyhydroxyl or complex, are considered flocculants or higher coagulants. The coagulation-flocculation polymer is clarified by the model scaffold, expecting that a polymer can stick to the surface of an ocean growth molecule with portions staying wide in the arrangements. These portions are then ready to tie to the empty locales of other algae particles, framing a three-dimensional floc cross section.

Algone microscopic fish a cell can be viewed as a little circular protest falling into a constant gooey liquid medium driven by constrain descending gravity and drag (or rubbing) and suspending powers. On the off chance that the molecule alga falls into the thick liquid from its weight, by gravity, at that point the max speed, otherwise called sedimentation speed, is achieved when the grinding power consolidated with the coasting power, these equilibria are precisely the gravitational power as portrayed in the law Of Stokes'. Indeed, planktonic sedimentation speed in wild algae is directed by an assortment of complex elements including cell portability, turbulence and water stream, and wind-incited increment and temperature stratification. The sedimentation rate of planktonic algae in a biological system can be lessened by the accompanying:

1. Motility
2. Lessening cell measurements
3. The augmentation of the drag powers as in the Scenedesmus species, which contain seta.
4. Lessening cell thickness, as in many blue algae, which contain gas vacuoles.

In this way, the sedimentation speed of algal cells can be expanded by expanding the extent of the cells:

This guideline applies to partition forms where synthetic algae coagulants are added to shape drops of huge algae that settle quickly in the reactor or on the base of the tub. In actuality, air bubbles, which can be joined to the effectively framed algae flocculants, significantly lessen the floc thickness floc by making a buoy on the glass. Expanding the seriousness of gravity will build the sedimentation speed of the algae cells, which is realistic by applying outward powers to the algae suspensions.

HARVESTING METHODS

Microalgae are photosynthetic microorganisms that are bounteous in nature and can develop in various conditions. Algal biomass can be utilized to deliver high included esteem items, for example, biofuels (biodiesel, ethanol, biogas and biohydrogen), creature sustain, human nourishment supplements, for example, vitamins A, B1, B2, B12, C, E, folic corrosive, Pantothenic corrosive and nicotinic corrosive), Omega 3 unsaturated fats (eicosapentaenoic corrosive (EPA), docosahexaenoic corrosive (DHA)), angle encourage, chlorophyll and other healthy skin items and hostile to maturing creams, against aggravation creams And regenerators for skin creams). Distinctive strains of microalgae contain a lot of protein (43-71% dry substance) practically identical to meat (43%), soybean (37%), drain (26%) and rice (8%). They deliver an extensive variety of amino acids that are fundamental for people and creatures that make extraordinary use in dietary supplements. Algae carbs (10-30% dry issue) are orchestrated as sugars, starch and polysaccharides (monosaccharides, disaccharides) which are anything but difficult to process. The oil content in the cells can represent 25-77% of the heaviness of the dry biomass. Microalgae are thought to be the best possibility for biodiesel generation since they are not in rivalry with the yield crops delivered by the best product lands (palm). These microalgae are delivered next to no time and give better execution on the market. Applying this idea in the biorefinery for the creation of biodiesel and other esteem added items to enhance the economy of biodiesel generation.

Be that as it may, the procedure of microalgae biodiesel and other esteem included items requires the development of microalgae, microalgae biomass recuperation which is vital for the creation and extraction and oil extraction process and other esteem included items. Be that as it may, the principle impediment in the utilization of microalgae biomass on a modern scale for the creation of significant worth included items is the lack of hydration step. Lack of hydration of microalgae around 20-30% of the aggregate cost of handling and creation of microalgae.

Microalgae products should think in light of the fact that there is a weakened suspension containing from 0.1 to 2.0 g of dry biomass per liter. The cost of removing and cleaning forms diminishes with the expanded convergence of biomass. In this manner, keeping in mind the end goal to get financially practical biodiesel generation, recuperation of microalgae ought to be less expensive. Different techniques for strong fluid detachment by drying or focusing microalgae societies of 10-450 g/l can be utilized. So the accompanying strategies were utilized:

- Sedimentation
- Vacuum filtration
- Cross stream filtration
- Pressure filtration
- Decanter centrifugation
- Disc stack centrifugation
- Dissolved air buoyancy
- Dispersed air buoyancy
- Micro bubble era
- Organic flocculation
- Inorganic flocculation
- Bio-flocculation
- Auto-flocculation
- Electrolytic coagulation
- Electrolytic flocculation
- Electrolytic buoyancy.

The point of this investigation was to exhibit current gathering strategies and grouping of microalgae and near examination to decide the most proficient and financially savvy drying out techniques for extensive scale creation of microalgae biomass for generation and biodiesel esteem included items.

These can be separated into two strategies:

- Physical Harvesting Method
- Chemical Harvesting Methods

Physical Harvesting Methods

Various physical strategies for microalgae drying out procedures have been utilized to recoup microalgae cells in its fluid suspension. These can be separated into four classes:

- Sedimentation
- Filtration
- Centrifugation
- Flotation.

Sedimentation

In this sedimentation strategy, the strong and fluid part is isolated from each other on the premise of the gravitational powers, as appeared in Figure 6.2. Different materials are isolated from each other in view of the molecule estimate and the thickness of the material.

A thickness contrast would bring about quicker sedimentation rates while a littler distinction in thickness and little particles would set aside a long opportunity to subside into longer gravitational powers.

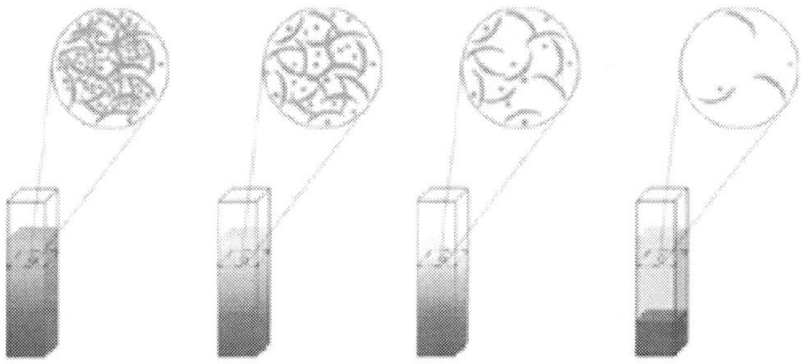

Figure 6.2: Sedimentation of microalgae

Type of Settling Tanks

There are two types of Settling Tanks:

- Lamella separator
- sedimentation tanks

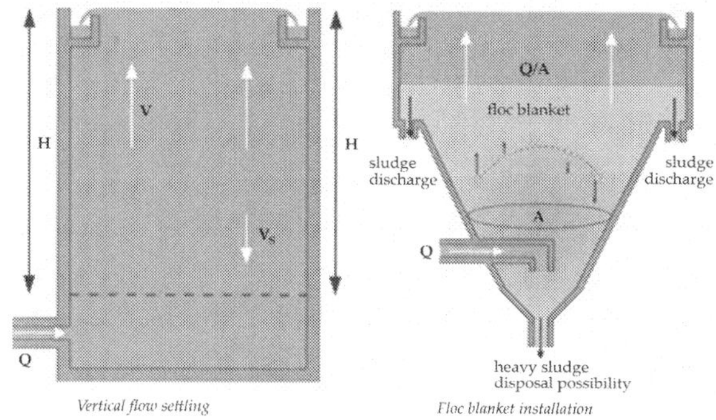

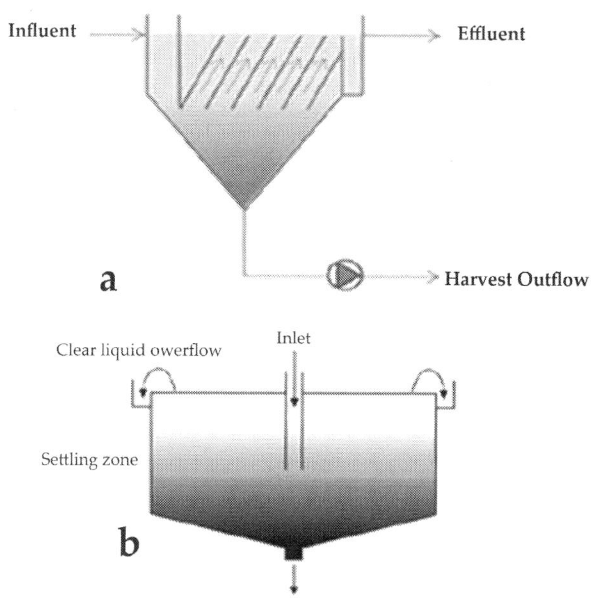

Figure 6.3: Sedimentation Tank (a) Lamella separator Tank (b) Generic Sedimentation Tank

SOURCE:https://www.slideshare.net/skpatil001/sedimentation-tank

Lamellar separators (Figure 6.3a) and sedimentation cradle (Figure 6-3b) are utilized to isolate solids from fluids.

- Lamella separators: They have higher traditional thickening range sedimentation because of the introduction of the plate. Lamella tanks work by embeddings the microalgal biomass through the delta. Fluid buoys at first glance (emanating) and biomass are caught on the slanted plates. After some time, biomass is saved on the base of the tub and is effortlessly gathered through the yield of the product. Process speed is quick a direct result of the vast sedimentation surface.

- Sedimentation: The tanks are round and hollow with a base channel so that the microalgae settled near the exit. The yield is situated at the base of the tank so the gathering of set up microalgae can be effortlessly recouped. The tank is outfitted with a pump that vehicles microalgae biomass from the way of life tank to the sedimentator through the passage. These supplies include that permits denser solids store on the base of the tub, leaving clear water at first glance. Once the sedimentation procedure has been finished, the algae can be recuperated from the tank through the yield.

Components Affecting Sedimentation

The variables affecting the settlement rates of microalgae include:

- Density and molecule estimate
- Temperature
- Aging of the cells
- Light power
- Time

Thickness and molecule estimate: The thickness of the microalgae is unique; microalgae change between 1030-1100 kg/m3 and the miniaturized scale algae freshwater thickness shifts between 1040-1140 kg/m3. Logical Granados et al. he detailed new water thickness (1000 kg/m3) and salt water (1025 kg/m3) are like those of microalgae and thus the sedimentation rate of algae particles is low. Murphy and Allen expressed that it is a test to dispose of microalgae biomass liquids because of the same cell and medium thickness. A researchfocused on the that the measure of extensive and little ocean growth sedimentation was 2.6 cm/h and under 1.0 cm/h. Cole and Buchak depict the sedimentation rate relies upon the kind of microalgae and found that green microalga has a normal sedimentation rate of 0.1 m/d. It has been watched that the sedimentation rate of 24 diverse microalgae species (with a size of 10 to 1000 microns) changes from 0.4 to 2.2 m/d and looks like cell size and fall rates. Milledge and sky uncover a sedimentation rate of 0.1 m/d for Chlorella in freshwater. Temperature:It has been demonstrated that the 4 ° C temperature settled an extensive variety of microalga species following 24 hours when the pH was balanced in the scope of 8-8.5.

Light: It has been notedd that prominent that the nonappearance of light has expanded the sedimentation rate amid high and low development stages. The supernatant got amid the high development stage contained 0.57 g/l of biomass within the sight of light and 0.39 g/l without light, while the supernatant got amid development start-up containing 0.28 g/l in theLight nearness and 0.17 g/l without light. It has been noticed that the convergence of microalgae in light and dull was 1075 cells/ml and 775 cells/ml, separately. Besides, it has demonstrated that sedimentation rates saw in nearness and nonappearance of light in calcitrans Chaetoceros were comparable.

Time: The convergence of microalgae sedimentation requires much sedimentation time to be more noteworthy than 24 hours. Stop and others. he watched long maintenance times of 1-2 days for recouping algae in a decantation shower scale. It has demonstrated that the rate increment and settling time 15 days expanded 94% sedimentation effectiveness. Scientists have discovered that the level of biomass recuperation following 24 hours of sedimentation S. platensis, C. fusiformis, T. suecica, Scenedesmus Nannochloropsis and was 95%, 96%, 80%, 59%, and 86%, individually.

Advantages and Disadvantages

- Sedimentation tanks are extremely compelling in the centralization of microalgae suspensions at 1.5% aggregate suspended solids (TSS); they are not broadly utilized at theindustrial level.
- Costs related with gravitational sedimentation are low, however the unwavering quality of this technique without the utilization of flocculants is additionally low.
- The microalgae required settling time is any longer than different procedures and the vitality is expected to pump the suspension.

- Ballesteros-González Fernández and said that this strategy is tedious and that cell sythesis can change.
- Cell focuses got by sedimentation are low.
- A gathering of microalgal biomass by single sedimentation is not the most effective technique since cell recuperation rates of 60-65% are low.

Filtration

The filtration algae gathering strategy is utilized as a porous medium that may contain algal biomass while permitting liquid section. This technique requires a weight distinction through the channel that can be driven by vacuum, weight and gravity. Film channels can be grouped by pore estimate in macrofiltration (more prominent than 10 microns), microfiltration (0.1-10 micron), ultrafiltration (0.02 to 0.20 micrometers) and turn around osmosis (under 0.001 microns). The weight required to constrain the liquid through declines the extent of the pores of the layer increments. Sifting procedures can amass microalgae cells in suspension 5 to 18% and working expenses shift from $ 10 to $ 20/lady. Filtration effectiveness strategies collected utilizing in the vicinity of 20% and 90%.

Vacuum Filtration

Vacuum filtration isolates solids from fluids by catching strong particles in a channel while pulling the fluid suction channel (Figure 6.4). Miniaturized scale algae measurements of 2 to 30 microns show that a microfiltration film is reasonable for vacuum filtration. Milledge and expressed sky that macrofiltration films can be utilized for extensive microalgae cells or algal cells are flocculation together. It has been accounted for that the vacuum filtration method is more reasonable for huge microalgae cells (more prominent than 10 microns). Spirulina platensis isolated species utilizing vacuum filtration outfitted with recovered cellulose film with a pore size of 0.45 microns.

Types of vacuum filters: There are five diverse channel films that can be utilized as a part of vacuum filtration.

They are:

1. Vacuum drum channel

2. Suction channel
3. Filter thickener
4. Belt channel
5. Starch pre-covered drum channel.
6. Mohn notes that the digestive suction channel, precooked channel drum, and sewing tape were appropriate to focus Coelastrum of microalgae in a scope of 5-37%. The creator found that drum channels were ineffectual gathering systems because of blockage. Thickening channel is not prescribed because of low solids content (3-7%) and high vitality necessities. A resaercher noticed that vacuum filtration outfitted with a Buchner channel and cellulose fiber channels were successful in recovering microalga arrangement.

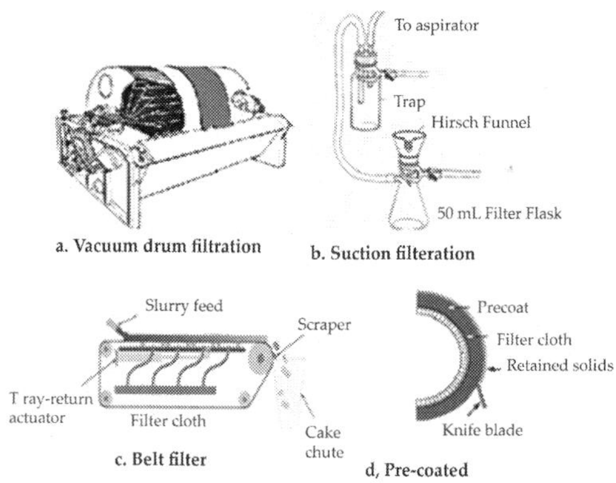

Figure 6.4: Types of vacuum filteration

SOURCE:http://www.solidliquid-separation.com/vacuumfilters/drum/drum.htm

7. There has been an effective recuperation of microalgae cells utilizing fossilized flour as a channel help to maintain a strategic distance from channel stopping up. Chaumont Gudin and announced that the filtration crate pre-weted with filtration help (diatomaceous earth) is successful in the accumulation of microalgae Chlamydomonas reinhardtii. Numerous specialists have assessed cellulose fiber channels and sand channels and those got inadmissible outcomes, however I consider the Earth

Dot Detector successfully recouped smaller scale measured Dunaliella species. Record for a surprising recuperation of the Dunaliella helped species utilizing a confirmed biatomic earth channel. Brennan and Owende expressed that the utilization of diatomaceous earth as a filtration help can viably expel the normal microalgae cells.

Energy consumption: Specialists have revealed vitality utilization in the range from 0.1 to 5.9 kWh/m3 relying upon the kind of channel utilized. Mohn said vacuum filtration expends 5.9 kWh of vitality/m3 for the grouping of solids suspended in arrangement 18-27%. They additionally announced that the vitality expected to dry proboscideum C with suction channel (8% SS), tape channel (9.5% SS) and channel thickener (5-7% SS) was 0.1, 0, 45 and 1.6 kWh/m3, individually. Umesh said fusiformis extraction smaller scale algae Spirulina through expansive pores was costly ($ 83.3/m2) and can gather 23 kg/m2 kWh. It has been accounted for by the analyst that the drop in weight for the vacuum procedure is in the scope of 70-80 kPa. Milledge and 0.25 kWh sky/m3 utilization for gathering microalgae utilized as a part of vacuum belt channel. Mohn announced power utilization of 3 kWh/m3 for vacuum filtration of microalgae.

Advantages and disadvantages: The benefits of utilizing the vacuum filtration procedure to gather microalgae are cell maintenance after the recuperation procedure. The productivity of the filtration procedure relies upon the span of the layer and the extent of the microalgae cells. An accumulation of microalgae by filtration is more productive than sedimentation, however the impediments of this technique incorporate layer substitution as well as occasional washing of the film to anticipate layer pores obstructing. In any case, the detriments are related with extensive vitality needs and expenses related with film substitution occasionally because of blockage. Plots and Collins recovered microalgae chlorophyll remove by vacuum filtration at 6 inches Hg (20 kPa) and high weights and noticed that drawn out filtration (more than 10 minutes) could harm the cells. It has been noticed that the filtration strategy is appropriate for bigger cell microalgae yet is not reasonable for recovery of microalgal species. The fast stopping up film was discovered utilizing the ultrafiltration layer procedure as verified by an analyst. Help filtration flocculation forms toreduce the requirement for vitality, however extra expenses for the woolly. It has recouped microalgae biomass utilizing the filtration technique and presumed that this accumulation strategy is not monetarily attainable for substantial scale generation.

Weight Filtration: Under weight, filtration is a procedure used to isolate the particles that frame a smaller fluid suspension. Weight filtration is utilized for isolating the fluid from the particles through the weight. The move through the channel is made by expanding the weight over the environmental weight to make a weight differential through the channel. This procedure is utilized as a part of the most ordinarily utilized parcel encourage and released into a persistent procedure. An autoclave upstream of the channel is required, and one is expected to gather the filtrate.

Sort of weight filtration: Gathering by weight filtration can be gotten with a channel plate and a press casing or utilizing a weight vessel outfitted with channels as appeared in Figure 6.5. The press channels plate and the edge constraining liquid into the suspension through the high-pressure microalgae channel. A progression of up close and personal rectangular plates are mounted vertically including the printing framework. A channel texture connected to each of the plates and held together by the power beneath the connected weight. The pumped liquid algae contained in the spaces between the plates and the weight is connected to compel liquid through the yields of the jars and the filtration plate. After division, the dried microalga cake is recuperated.

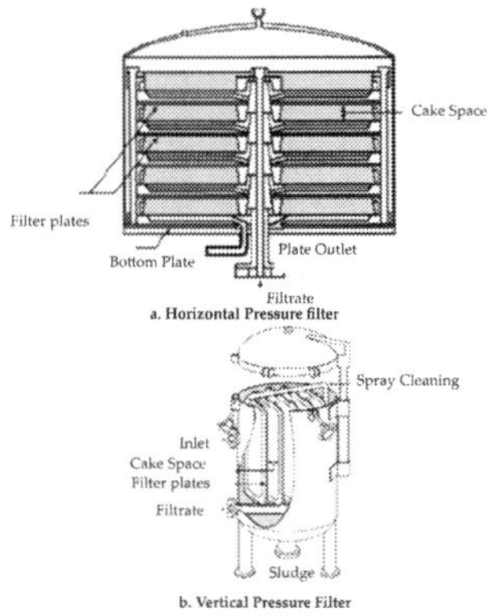

Figure 6.5: *Pressure Filter*

SOURCE: https://www.dicalite-europe.com/filtrat.htm

Energy consumption: This technique can be considered vitality effective in light of the fact that a base measure of vitality is expected to assess item generation and the additional amount of crude material begins. Be that as it may, the viability of the technique relies upon the kind of algae.The measure of vitality expected to drive 22 to 27% (w/v) of C.paroboscideum species utilizing weight channels is 0.88kWh/m3. Nagle and Lemke watched a convergence of microalgae at 8% (0.5%) utilizing a press channel having 20 plates and a body (30 cm measurement) of channel paper having a pore size of 5 microns. It has been noticed that microalga Dunaliella Chlorella was too little to ever be recouped for low-weight filtration. Mohn revealed that the vitality devoured for gathering C. round and hollow proboscidean (7.5% suspended solids focus) and channel wicker bin (5% of suspended solids fixation) was 0.3kWh/m3 and 0, 2KWH/m3,respectively. It was likewise discovered that the weight filtration was not appropriate for Scenedesmus, Chlorella and Dunaliella species, but rather it was attractive for other microalga species as expansive proboscideum Coelastrum and Spirulina platensis.

Points of interest and weaknesses: A few points of interest of utilizing crisp filtration are gathered (compound particles in fluid suspension) having a low dampness content, dissolvable cake recuperation is high, the channel for 1-2 minutes channel cleaning can accomplish high straightforwardness arrangements And materials can be utilized to assemble channels and manufactured amalgams and inward parts.

The inconveniences of utilizing this strategy are hard to wash the channel medium increments when the strong is sticky, within is hard to clean in sustenance applications and it is hard to see the state of the channel because of embodiment of the compartment.

Cross Flow Filtration

Vast volume volumetric microalgae gathering cells can be made effective utilizing digressive filtration, as appeared in Figure 6.6. In this method, the specimen streams extraneously through a layer. The molecule estimate not higher than the layer pores is kept up and returned as held. Littler particles go through the layer with the fluid and are called saturated.

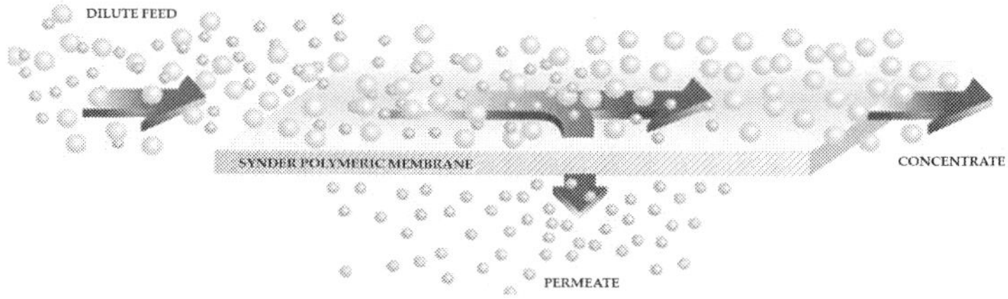

Figure 6.6: *Cross Flow Filtration*

Source: http://desalite.com/ultrafiltration.html

Membrane type: Ultrafiltration or microporous films are the sorts of channel layers utilized as a part of this strategy. These layers are accessible with an extensive variety of pore estimate and atomic weight maintenance. Polymeric layers have a long life expectancy when utilized as a part of proper speed and cross stream conditions at low-weight film. A scientist utilized a cross-filtration with a pore size of 0.45-micron film and accomplishes 70-89% recuperation productivity of biomass. It has been discovered that polymeric films are viable in recouping Skeletonema costatum types of microalgae and Haslea clam, yet yields rely upon hydrodynamic conditions, microalgae properties, and microalgal cell focus. It has been accounted for that the underlying stream for microfiltration. The layers have been far better than those of ultrafiltration, yet are all the more effectively obstructed. Zhang et al. he utilized a ultrafiltration film with cross-stream cross-stream of 0.17 m/s and watched an expansion in theconcentration of 0.104% to 92.5% of algae on the surface of the layer with a yield estimation of 46.01 g/ m2/h. 14 distinctive inorganic films have been tried with the ATZ-50 kDa ultrafiltration layer delineated by enhanced yields and thought Arthrospira plate species, a factor of 20. An extraneous filtration procedure with ultrafiltration natural film (polyacrylonitrile, 40 kDa) and Arthrospira focus plates a factor 10. Species Skeletonema costatumconcentrate utilizing ultrafiltration stream rate of 30 liters/h for 12h.In reality, they focused Dunaliella's cross-linkedultra-filtration saline at 30-40 l/h. It is the kind of Thalassiosira pseudonane concentrated 2, 3 L 2840L comprising of 2,33x1012 cells/l utilizing a microfiltration layer framework. It has been watched that cross-stream microfiltration cross-stream resistance as the cross-stream speed expanded from 0.13 to 4 m/s while Chlorella sp. species.

Vitality utilization: A scientist detailed that vitality utilization through theflow through stream filtration strategies can fluctuate from 3kWh/m3 to 10kWh/m3 relying upon electrical qualities, framework plan and connected weight. It has additionally been noticed that unrelated filtration can likewise be utilized as a part of powerless suspensions and is a helpful strategy for amassing suspended solids in the range from 2.5 to 8.9% with vitality in the scope of 0.38 to 2, 06kWh/m3. An examination announced that the vitality utilization of the extraneous filtration system was 5 kWh/m3.

Centrifugation

This kind of evacuation system is generally utilized as a part of drink, nourishment and pharmaceutical businesses. Centrifugation is a procedure in which a diffusive power is utilized to enhance detachment of solids. Turning the suspension makes the weight contrast expected to isolate the particles from the fluid suspension. In this manner, the proficiency of the recuperation procedure relies upon diffusive power.

Types of Centrifuges

The two sorts of centrifugation utilized for reaping microalgae are: circle stack and decanter axes.

- Disc stack axis
- Decanter axis

Disk stack axis: The most well-known modern axis is as of now utilized as a part of business offices for the generation of high-esteem items and algae biofuel is an outward bundle plate. It comprises of a base dish with an extensive number of heaps of divided metal cones (circles), as appeared in Figure 6.7 Separate Density.

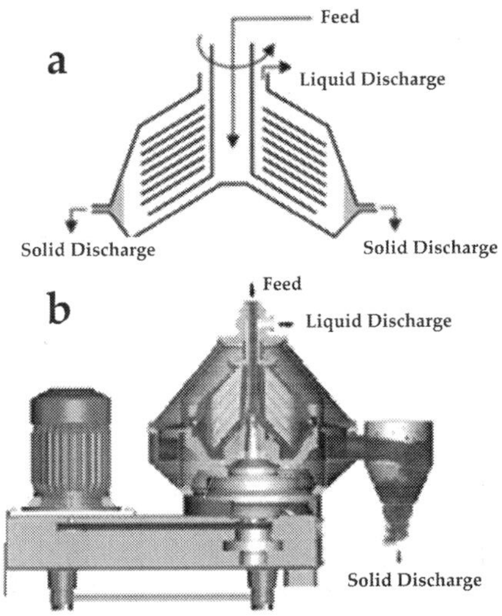

Figure 6.7: Disc Stack Centrifuge **(a)** Borowitzka and Mohelmani **(b)** Envirpre

SOURCE:http://longshortreport.com/2017/01/13/global-disk-stack-centrifuge-market-2017-2022/

The mix is set at the focal point of a pack and the lightest period of the mix remains focused, while the overwhelming stage is moved to the base of the plate. This strategy isolates the diverse layers thickness materials. It is more appropriate to isolate from a molecule estimate in the scope of 3 to 30 microns and suspensions with centralizations of solids content in the vicinity of 2 and 25% of particles.

Recuperation cell productivity nine distinct types of microalgae utilizing a bundled rotator assessed and more noteworthy adequacy 95 to 13,000 g mending was watched. They additionally found that recuperation effectiveness diminished with 60% lessening in gravity and 40% with gravitational powers 6000 and 1300 g, individually. Specialists noted 90% effectiveness microalgae expulsion utilizing a radiating sharp edge sort. A convergence of 18% of microalgae utilizing a plate axis is come to. Mackay utilized a plate rotator working at a weight of 4000 to 15,000 g of a suspension containing 0.2 to 20% biomass% v/v of algal cells. Spirulina plug sp. utilizing an outward drive sort keeps running at 6000 rpm for 5 min.

Decanter rotator: Decantation by centrifugation depends on the idea of utilizing an extraordinary sedimentation tank where suspended solids are constrained because of gravitational powers. The decanter (Figure 6.8) runs constantly pumping the microalgal biomass developed in a can axis with the goal that the particles suspended in the arrangement are compelled to the base of the cuvette. The staying fluid after the molecule partition goes through the flood tube.

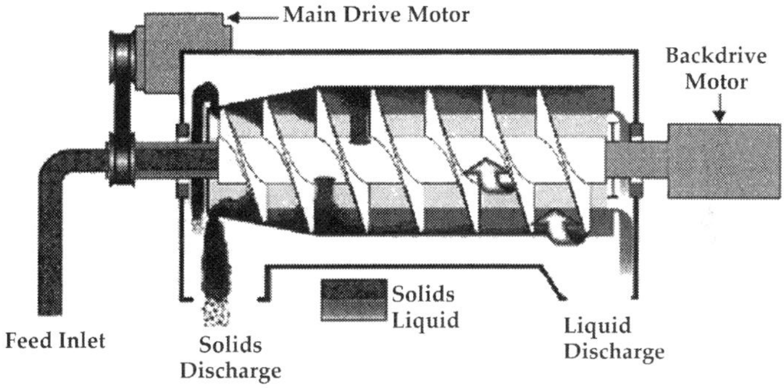

Figure 6.8: Decanter centrifuge

SOURCE:http://www.gn-decanter-centrifuge.com/mud/decanter-centrifuge-working-principle-decanter-bowl

The centralization of microalgae biomass utilizing a decanter axis is superior to other gathering techniques. Smith and Paper expressed that the immaculateness of the fluid delivered after partition was not as incredible as that got by centrifuging the holder. Mackay decanter utilizes a resistance of 4000-10000 g for slime containing 5-80% biomass. Biomass microalgae must be subjected to an underlying thickening stage with buoyancy air disintegrated to focus the suspension of microalgae (0.02-0.05% by weight) 2-3% preceding by decanter centrifugation.

Vitality Consumption

Outward Packaging: The vitality utilization detailed in the writing for the turn rotator differs from 0.53kWh/m3 to 5.5kWh/m3. Alfa Laval utilized a rotator for microalga dry lack of hydration sort, making 16% of the 0.53kWh/m3 vitality utilization. Mohn watched a 12% strong suspension of the microscenescenedesmus species utilizing a circle axis with theenergy of 1 kWh/m3. Goh collected pigsmicroalgae squander developed with radiating utilization 1.4kWh/m3. A resaecrher noticed that the 5.5kWh axis bundled/m3 devoured Chlorella sp.

Preferences and Disadvantages

Disk-stack axis: The benefit of utilizing slinger stockpiling to gather microalgae is its high expulsion effectiveness contrasted with other modern axes. The sustain focus for these units is normally in the scope of 0.5 to 10% w/w. This kind of axis oversees expansive volumes and can isolate fine particles (from 0.1 to 100 microns). This gadget can be utilized to isolate the strong from the consistent cluster liquid, semi operation. A few burdens of this kind of axis include: low mechanical dry and dry substance of the mechanically complex seepage framework and little space between the stacked circles turns out to be harder to clean and may require cleaning chemicals.

Decanter rotator: Got dried out biomass utilizing decanter is considerably more thought than that got with the circle axis. In any case, the decanter is more reasonable for strong suspensions with higher particulate suspensions and is not appropriate for microalgae. This sort of axis is most appropriate for division of particles bigger than 15 microns and strong suspensions containing over 15% of materials. It works with resistances underneath 6000 g of inactivity. The hindrances of this strategy for gathering microalgae are exceptionally focused power required (normally inside the scope of 4-40% w/w), the active liquid the framework cannot be clear because of the nearness of fine, delicate solids thinning Thickened and cause mechanical troubles, substantially more vitality than plate rotators and the high expenses related with the hardware expected to process expansive volumes. It was assessed that this sort of axis expends 3000 kWh/ton of dry marine biomass.

Buoyancy

Portable innovation is utilized to gather low microalga safe densities. This strategy is delegated a concoction physical gravimetric detachment process in which gas bubbles go through a strong fluid from the buoy to the miniaturized scale algae surface gurgling suspension. Ventilation likewise helps evacuate the unpredictable natural mixes show in the water arrangement gave cleaner download. The adequacy of this technique relies upon the precariousness of suspended particles, the more noteworthy the contact with the air particles is low instability. In buoyancy strategy, the measure of the particles is not essential, the littler the molecule estimate is the more inclined to raise base air pockets. Arrangements with particles littler than 500 microns can be utilized as a part of buoys.

Types of Air Flotation

The buoyancy forms are gathered by the technique that is utilized for the air pocket arrangement into:

- Dispersed air buoyancy
- Dissolved air buoyancy

Scattered air buoyancy: This method requires the utilization of a rapid mechanical stirrer for bubble development and an air infusion framework as appeared in Figure 6.9. The gas is blended with the fluid this is presented at the highest point of the compartment and can pass an air pocket sink going from 700 to 1500 microns. These air pockets are bigger (1000 micron in breadth) than that of the item utilizing separated air buoyancy strategies. Chen et al. utilized scattered air buoyancy strategy for drying out of Scenedesmus quadricauda bowing in air microalgae. They utilized three surfactant factors to evacuate cell microalgae:

- Nonionic X-100
- N-N-N-N-trimethylammonium bromide N-cetyl and
- Sodium dodecyl anionic sulfate

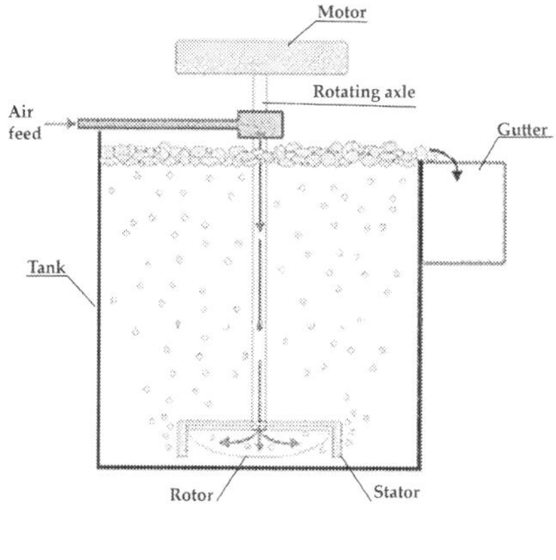

Figure 6.9: A diagram for Dispersed air-flotation

They additionally found that surfactants assumed a vital part in expanding the honesty by abstaining from percolating and found that this expulsion technique is best with the utilization of N-N-N-N-N-N-trimethylammonium N-cetyl surfactants.

Yan and Jameson noticed that the air-scattered air innovation included dispensing with 98% of the viability of microalgae. Xu et al announced a recuperation effectiveness of 93.6%.

Broken up air buoyancy: The method of buoyancy air-terminated buoyancy is most normally utilized as a part of the treatment of mechanical effluents. This technique requires a lessening in the water weight that is soaked with the air. The fluid is then infused into the buoyancy tank at barometrical weight. Air pockets are created by the spouts of the diffuser and ascend through the microalgae cells conveying fluid suspended by and large for the surface of the tanks as appeared in Figure 6.10. The biomass aggregated at first glance can be skimmed and reaped. The elucidated fluid segment is immersed with air and reused to the buoyancy tank. The air supply framework can be controlled by shifting the immersion weight or by changing the reused extent to the tank. The extent of the framed air pockets can be controlled by the saturator being overwhelmed by environmental weight and the infusion rate. The stream rate must be adequately expansive to anticipate reflux, giving a weight drop and enabling rises to develop on the surface of the tubes. Little air pockets are attractive with a size of 10 to 100 microns (with a normal size of 40 micrometers).

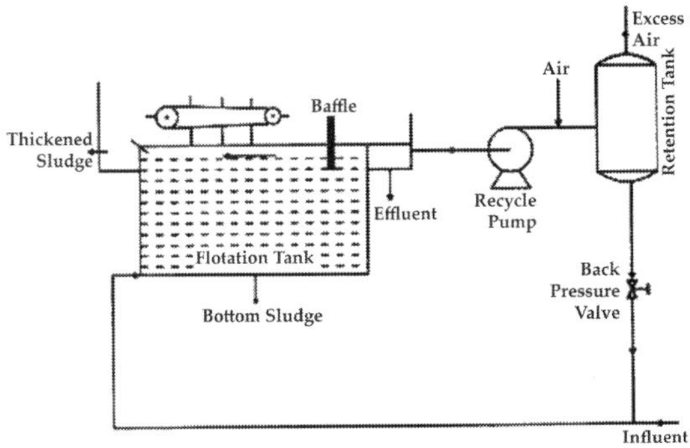

Figure 6.10: Apparatus for **air-flotation**

SOURCE. http://www.yourarticlelibrary.com/essay/flotation-as-primary-treatment-of-waste-water-explained-with-diagram/28341/

Separated partition of microalga in theair combined with the normal utilization of a synthetic flocculation process. Edzwald considered the utilization of microalgae for skimming air recuperation and found that the pretreatment strategy required for flocculation however was more fruitful than sedimentation system. Released air buoyancy was utilized to gather and decide 5% microalgaesuspended solids with an utilization of 7.6kWh/m3. Goh confirmed that this technique is powerful to collect microalgae pig suspension when the buoy joined with a high measurements of 0.3 g/l.

- Fluidic wavering: Fluid drifting microalga skim evacuation expels the singular utilization of a compound flocculation process. Edzwald was examined utilizing microalgae disintegration analyzation in air disintegration and this was seen with pretreatment technique for pre-flocculation yet was risotto sedimentation innovation. Disodium wind stream utilized for raclette and microalgae strong outcomes. A concentrated suspension of 5% of vitality utilization is 7.6 kWh/m3. Goh reveal this successful microalgae gathering technique gives swine pig with regards to a mix with a high substance of 0.3 g/l,

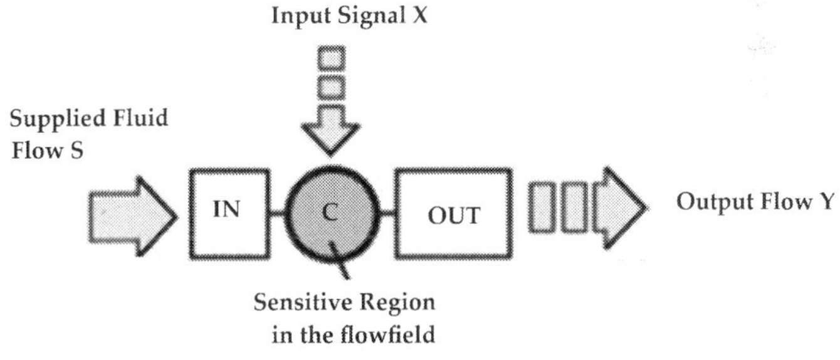

Figure 6.11: Rule of fluidic wavering the adjustment in yield stream (Y) is caused by the liquid's ceaseless stream (S)

They revealed the recuperation productivity of 99.2% microalgae by creating microbubbles at pH 5 with the guide of ferric coagulant coagulation (150 mg/l). More productivity was seen than 95% of the disposal of skimming microbubbles. The expulsion proficiency was seen more than 95% for the evacuation of Microcystis and Cylindrospermospsis fibers utilizing gliding microbubbles. Nonetheless, very few investigates have been finished utilizing this microalga recuperation system to be considered monetarily reasonable for substantial scale recovery of microalgae cells.

Factors Affecting Air Flotation

Elements that impact the proficiency of microalgae gathered utilizing the air buoyancy method include: pH, wind stream, alkalinity, distribution rate, pressure driven load and time. Skimmer speed decides the convergence of the organized suspension and tallness over the surface of the water.

pH: A standout amongst the most imperative parameters that influence buoyancy forms is pH. Lin and Liu said pH influences response ways and interface properties. Keeping up the pH in the 5-8 territory with sodium dodecyl sulfate has next to zero impact on the expulsion productivity (95%) of quadricycle scenicusmus microalgae. They ascribed this to electrostatic communications between the microalgae surface and surfactant surface gave by pH esteems underneath 8. Notwithstanding, at pH more than 8 the collaboration between the surface of algae and surfactants was feeble; he didn't permit the cancellation. Kwon et al. detailed that the buoyancy flocculation strategy required air pH 7-8 and 5-6 flocculant organic for inorganic flocculants to accomplish evacuation productivity more than 90%.

Alkalinity: A standout amongst the most vital parameters is pH control. Lin and Liu clarified that pH influences response ways and interface properties. They ascribe this to electrostatic connections on the surface of the positive microalgae and the surfactant is given by pH esteems beneath 8. It has been accounted for that the drifting air natural pH 7-8 and 5-6 for inorganic flocculants is required accomplish expulsion efficiencies more than 90%.

Wind stream rate: Concentrates in the writing demonstrate that the variety in the wind stream rate of 68-206 ml/min did not influence the expulsion proficiency (92%) of the Chlorella sp. Utilizing the buoyancy method at a speed of 100 L/h. Dassey and Theegala announced that an expansion in wind current past 160 ml/min does not expand Significantly the generation of microbubbles and all things considered no increment was seen in gathering effectiveness, Hanotu and others noticed that the wind current rate of 85 l/min was powerful in isolating half-microalgae.

Medium: Analysts watched that algal cells were denied of nitrogen had bring down centralizations of surface utilitarian gatherings with the exponential stride of stationary development and decay. They have discovered that these useful gatherings are framed by proteins and polysaccharides that are essential to balance out surface charge impacts bubble adherence in the purified air buoyancy procedure. It has likewise been watched that more coagulant alum is expected to accomplish more noteworthy gathering productivity (90%) in exponential stage contrasted with stationary and diminishing stages, as it sets aside greater opportunity to destabilize the surface of cells created in the exponential stage. There was a straight connection between's the convergence of practical surface gatherings and the alumination measurements for a given reaping productivity. They have demonstrated that cells with more water-gathering effectiveness have been watched for high development because of a higher zeta potential. Sustenance limitation on the media may enhance gathering proficiency of microalga buoyancy techniques.

Recycle rate: The reusing rate revealed in the writing went from 5 to 10%. Edzwald and Wingler got 97-99% expulsion effectiveness of Chlorella vulgaris species with a 8% reusing rate. Teixeira and Rosa detailed recuperation effectiveness 92-98% of blue-green microalgae utilizing a reusing rate of 8% and said reusing framework was crucial for successful molecule recuperation yet reuse rates surpassed 8%. They additionally found that the option of pressurized reusing framework did not enhance the recuperation rate of the cells. This marvel is credited to the absence of molecule destabilization since molecule destabilization is basic to the adequacy of air-scattered air contrary to the measure of the piece.

Hydraulic stacking: The mechanical water powered load rate buoyancy edge applications fluctuate from 0504 m/h to 40 m/h. Edzwald detailed rapid released wind current methods can be performed at water driven heaps of 20-40 m/h. Specialists noticed that a more prominent pressure driven load lessens the development of air pockets. It has been accounted for that an expansion in pressure driven burdens diminished air disintegration time, which brought about poor air pocket yield.

Time: The time required for the real wind current of air announced in the writing went from 3 to 30 minutes. Edzwald and Wingler noticed that the time required expelling 96-99% of Chlorella vulgaris via air buoyancy was 5 min. An air buoyancy method to think Microcystis aeruginosa species to 94.5% of every 8 min was utilized by the scientists. It increased 80% evacuation effectiveness of Cyclotella Melosira for air buoyancy in 3 min. It has been watched a recuperation effectiveness of 93.6% for B. braunii utilizing air buoyancy at 14 min. Weakling et al. He watched that the air gliding method viably microalga gathered inside 30 min.

Vitality Consumption

Buoyancy method is a high vitality process because of the high weight required. Algae collecting utilizing disjoined air buoyancy is a powerful strategy yet with high working expenses related with the utilization of a serious vitality compressor working at 390 kPa weights. Be that as it may, the fluidic oscillator does not require much vitality to work and has been seen to devour 2-3 arranges less vitality than the released and scattered air buoyancy strategies.

Points of interest and Disadvantages

Contrasted with sedimentation, the buoyancy strategy is considerably speedier and more viable gathering microalgae. Mohn said microalgae lack of hydration with buoyancy techniques (7% focus) is significantly quicker and more effective than sedimentation (fixation 1.5%). Be that as it may, the buoyancy technique has just been accounted for to be powerful in microalgae gathering at a bank scale and is not reasonable for the recuperation of extensive scale microalgae cells. Furthermore, this technique is serious vitality, it has high working expenses and high vitality is required for the little air pocket era.

Concoction Harvesting Methods

In light of vitality necessities, concoction flocculation and drying out strategy is by all accounts the most encouraging for extensive scale utilize.

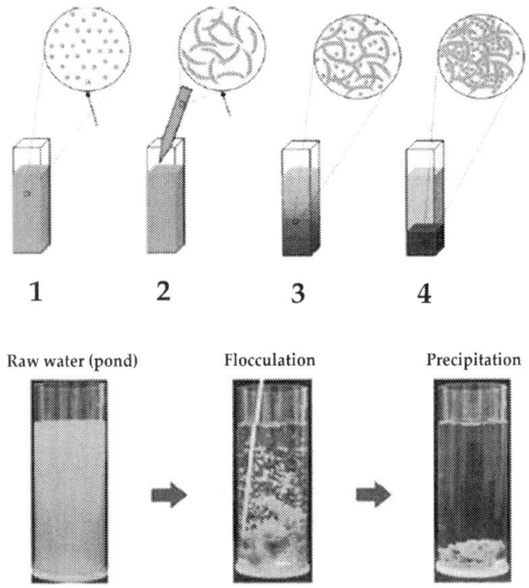

Figure 6.12: Flocculating harvesting process of microalgae

Source: http://energyeducation.ca/encyclopedia/Harvesting_algae

These techniques work to focus flocculation cells (coagulation) trailed by settlement in the lower some portion of the product in view of the higher thickness of the pack as appeared in Figure 6.12.

Sort of Chemical Flocculation

There are two sorts of chemicals (Table 6.1) that can actuate flocculation:

- Inorganic polymers.
- Organic polymers.

The impacts of these polymers on microalgae are appeared in Table 6.2. Regularly, cationic, anionic and non-ionic polyelectrolytes are utilized to flocculate the microalgae cells:

Table 6.1 Examples of different chemical flocculent

Chemical Type	Flocculent
Inorganic	Ferric sulphate
	Ferric chloride
	Aluminium chloride
	Aluminium sulfate
Organic	Magnafloc LT 25
	Zetag
	Praestol
	Chitosan

Table 6.2: Effect of different chemical flocculent on microalgae

Flocculent Type	Effect on Microalgae
Inorganic	Toxic to the cells
	Alter the color of the medium
	Alter the pH of the media which may make it unsuitable for reuse
Organic	High cell viability (>75%)
	No inhibitory effect the cells
	Non-toxic

Inorganic Compounds

They have achieved Chlorella species with a 85% expansion in collecting productivity with press salts. Analysts have detailed pooling proficiency of over 90% for Chlorella zofingiensis species utilizing a ferric chloride centralization of 40% (w/v) at pH 4.0. Xuan expulsion productivity achieved 90% for Nannochloropsis sp. Utilizing ferric chloride managed at 0.18 mg/l. An exploration has announced accomplishing over 80% proficiency for marine flocculation microalgae with ferric chloride. Bintisaarani found that ferric chloride flocculation was the best to harvest and recuperating Nannochloropsis 89% species evacuation productivity utilizing a ferric chloride grouping of 0.9 M at pH 7.5. The salts were seen to successfully skim Chlorella microalgae and Scenedesmus aluminum species. Another examination found that aluminum salts were more successful in woolly Chlorellas species than press salts. It has been watched that flocculant alum was better than ferric sulfate as far as pH, measure of flocculant, and nature of the last suspension of water.

Natural Compounds

Flocculating natural polymers (chitosan) or polyelectrolytes (polyelectrolyte polyamine) are known as polymeric flocculants (manufactured and regular) comprising of ionic and non-ionic species. The utilization of flocculation natural mixes works by joining both extension particles and charge balance. The charge thickness and length of the polymer bind decides the degree to which each is utilized. The procedure begins with polymer assault on the surface of microalga by synthetic or electrostatic powers.

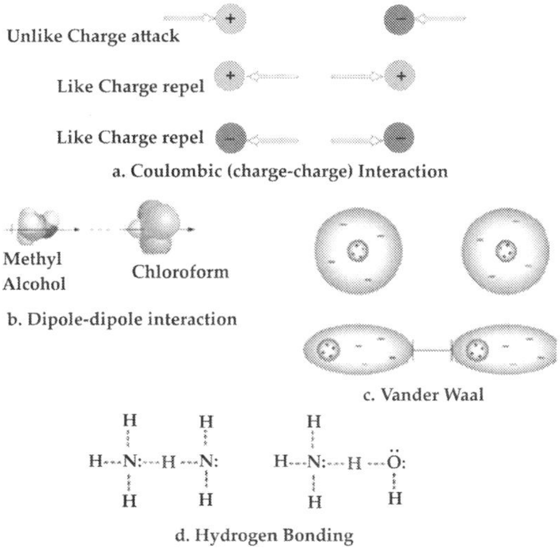

Figure 6.13: Ways to attach a polymer to the cell surface

The polymer can tie to the cell surface through Coulombian collaboration (stack stacking), dipole, van der Waals or hydrogen as appeared in Figure 6.13. The Coulmbic fascination constrain acts with various charges on the polymeric surface and microalgae are consolidated, after the idea that equivalent charges are commonly quieted and, diverse charges draw in. Dipole-dipole cooperations happen when two particles approach polar and incompletely negative parts combine the halfway positive atom. The powers of Van der Waals are powers of intermolecular fascination between atoms. Hydrogen holding is a sort of dipole fascination where a hydrogen particle is bound to an exceptionally electronegative molecule. Accordingly, the polymer ties to the surface leaving the tail in solvents shaping circles. Circle and polymer lines enable attaching to other crossing over cells between them.

The productivity of this procedure relies upon the level of flocculation in microalgae cells that cover the polymer. On the off chance that the polymer assault on the surface of the cell is not as much as the ideal sum, at that point it will be unable to withstand slicing powers because of tumult. Likewise, abundance polymer covering on cell surface can cause static connecting impediment process.

Late examinations have uncovered that cationic flocculant polyelectrolyte operators are more successful for microalgae recuperation. It was noticed that inorganic flocculants were less effective than flocculation Muriellopsis sp. different sorts of natural specialists. It was accounted for that the expansion of 0.5 mg/l of non-ionic polymers Magnafloc LT25 (anionic polyacrylamide BASF substance organization) was set up on a medium with a pH changed in accordance with 10 to 10.6 and adequately focusing an extensive variety of animal categories Microalgae at rates that are 200-800 times higher than that of control. Harith et al. demonstrated that they kept up high cell essentialness of microalgae (75%) for Chaetoceros species utilizing Magnafloc flocculant LT25 at a measurement of 1 mg/L. They said that the expansion in measurements of Magnafloc LT 25 and Magnafloc LT 27 did not expand the viability of Flocculation but rather an expansion in sedimentation rate.

T'Lam et al. revealed flocculation proficiency of 98% for P. tricorntum utilizing flocculant zetag 10 ppm however just achieved 52% was gotten for oleoabundans N species. Udom et al. he found that the utilization of zetag at the measurements of 34 mg/l of flocculated microalgae and brought about a recuperation effectiveness of 98%. Buelna et al. he watched a 95-100% evacuation effectiveness to develop Chlorella utilizing zetag 63 5 mg/L at pH 9.6.

Pushparaj et al. Teraselmis and Spirulina flocculation with a recuperation productivity of 70% by utilizing biomass Praestol (flocculant in view of cationic polyacrylamide natural) without inhibitory consequences for reused and reused media. Notwithstanding, there has been flocculation in cationic natural polymers in rooms with saltiness higher than 5 g/l. It has been discovered that the measure of flocculant expected to execute 90% of the microalgae of the fluid suspension expanded straightly with more noteworthy saltiness. Scientists noticed that the measure of vitality required acquiring a 15% (w/v) microalgae fixation utilizing flocculation polymer was 14.81kWh/m3.

Chen et al. he announced that general chitosan measurement interim required flocculant species viably microalgae was 5-200 mg/L. Xu et al. watched 99% adequacy for clarithromycin sorbitan chitosan with an ideal measurements of 10 mg/g of microalgae and pH esteems dried less than 7. Ahmed et al. It achieved a 99% yield of flocculants 20 minutes for Chlorella sp. Chang and Lee, submitted 99% productivity for vulgaris Chlorella flocculation utilizing chitosan at a measurements of 200 mg/l and a pH of 8.7. Sirin et al. he announced a flocculation proficiency of 92% out of 10 min for every Phaeodactylum species utilizing Chitosan (20 mg/L). Spirits et al. 100% flocculation effectiveness was watched Chlorella sp. Utilizing chitosan at a convergence of 40 mg/l. Shoreline et al. They looked at the techniques for flocculation, centrifugation and chitosan filtration for microalgae accumulation and flocculation saw that chitosan was the strategy devours less vitality than the three.

Factors Affecting Chemical Flocculation:

Inorganic Flocculation: Components that impact inorganic flocculation incorporate flocculant fixation, pH, and flocculant shallow charge.

Woolly focus: It has been watched that the fixation to which the flocculant is controlled in the framework influences the viability of microalga recuperation. Rakesh et al. utilized groupings of aluminum sulfate from 50 to 300 mg/L for the recuperation of Chlorella sp., Chlorococcum sp and Chlorella sorokiniana and found that 50 mg/L was the best measurement. Garzon-Sanabria et al. Assessed recuperation productivity was figured utilizing aluminum chloride in focuses in the scope of 50-100 mg/L and found that 50 mg/L was the best measurements. Ferriols and Aguilar announced the utilization of calcium chloride and sodium hydroxide at a grouping of 100-200 mg/L for the recuperation of Tetraselmis terrahele and achieved a higher recuperation effectiveness of 200 mg/L. Scientists watched that in medium with low convergences of algae, the centralization of ferric chloride flocculated builds Chlorella zofingiensis directly with the cell fixation.

PH: Changing the pH of the medium utilizing inorganic flocculants can advance cell accumulation. Knuckey et al. Successful flocculation has been watched (> 80%) calcitrans Chaetoceros of Chlorella muelleri, Thalassiosira pseudonana, Attheya septentrionalis, Closterium Nitzschia, Skeletonema sp., Tetraselmis suecica Rhodomonas salina and changing the pH by including sodium hydroxide. Garzon-Sanabria et al. utilized aluminum chloride (50-100 mg/L) to alter pH (4-7) and accomplished better recuperation productivity by utilizing a 50 mg/L (pH = 5.3) recuperation. Lee et al. He noticed that the variety in sodium hydroxide fixation impacts Botryococcus braunii flocculation productivity because of pH variety in the medium. Sanyano et al. Chlorella sp. effectively rushed.

Surface charge: Shallow microalgae cells are stacked contrarily, demonstrating that a positive flocculant to tie the cells together would be required. Coagulation algae is instigated by the emphatically charged flocculant fascination on the phone surface adversely charged and official of another algal cell on the positive flocculant. The flocculation proficiency relies upon the measure of flocculant accessible to joint the algae together. It has been watched that the energy of ferric chloride incited flocculation microalgae with a recuperation proficiency of 90% at a pH higher than 4.1 and beneath 8. Knuckey et al. depicted overflowed microalgae with 80% proficiency utilizing Fe3 + particles. Washed down Nannocloris Recovered with 90% effectiveness utilizing aluminum chloride to check the shallow charge of microalgae cells at pH 5.3. Chlorella sp. effectively ran, Using ferric chloride. Lee and others got flocculation proficiency 100% Chlorella sp. Utilizing natural mixes of aluminum and magnesium incorporated.

Natural Flocculation: Components that impact natural flocculation include: pH, stack on polymer measurements and saltiness.

PH: A few types of microalgae can flocculate together by altering the pH. It has been expressed that the pH of the synthetic piece and the microalga media impact the measure of flocculant required. They watched electrostatic shock between low colloids at low pH, bringing about more noteworthy measure of extensions on the grounds that the polymer chains are longer. They likewise found that the measure of polymeric flocculant expected to shift the convergence of microalgae, on the grounds that the heap on the surface of the algae. It has been accounted for that the non-ionic Magnafloc LT25 polymer viably understood a wide assortment of microalga species utilizing a rate of 0.5 mg/l in a medium balanced at pH 10 to 10.6. Scientist found that the best flocculation was when cationic polyelectrolytes were utilized at low pH levels. It has been watched that Chlorella species flocculation when the pH expanded to 11-12. Outrageous pH levels can cause cell passing or disintegration.

Charge on polymer: Charge Polyelectrolyte assumes a critical part during the time spent flocculation of microalgae. Anionic polyelectrolytes are not successful flocculants because of the surface of adversely charged microalga cells as charges are rejected and/or the length of the polymer is insufficient to tie the particles together. It has in this manner rose that cationic polyelectrolyte is considerably more powerful in microalga flocculants. Morrissey et al. he noticed that the N, N-dimethylaminopropyl acrylamide (Positivity) polymer has brought about microalgae recuperation efficiencies more prominent than 99% at pH 7 and expanded pH at 13 (enactment of the negative capacity) given the productivity are flocculation under 12%. The chitosan surface decidedly charged 99% evacuation proficiency for Clorella vulgaris.

Woolly fixation: The measure of cationic flocculant required for powerful scaffold between the cells relies upon the measure of negative charge show in the framework, the surface charge thickness, cell tally in volume, cell surface and aggregate charge thickness decidedly charged polyelectrolyte. Negative charge at first glance is instigated by utilitarian gatherings (carboxy gatherings) in microalgae cells that have been seen to impact the isoelectric purpose of the phones. The development stage and metabolic states of microalgae cells decide the focus and reactivity of these practical gatherings. Granados et al. It has been demonstrated that Chlorella and Scenedesmus species were viably flocculated at a convergence of 2 g/L after 15 min utilizing polyelectrolytes 2-25 mg/g. The cationic polyelectrolyte polyamine was powerful in flocculant microalgae cells at a dosage of 2.5 mg/l. It has been accounted for that microalgae microflora Chlorella and flocculant stigmatophora galbana Isochrysis require 5-10 times more than those required by the dosage of freshwater microalgae.

Saltiness: The level of saltiness impacts the natural flocculation of microalgae. Bilanovic and Shelef noticed that the flocculant polyelectrolyte was hindered in the marine condition because of its high saltiness and compelling flocculation watched saltiness under 5 g/l. This has been credited to the way that the high ionic power makes the polymer shrivel, with the goal that the scaffold is not shaped to attach the microalgae cells. It has been accounted for that the expansion of antacid to Chlamydomonas did not create quick flocculation in saline.

Focal points and weaknesses

Contrasted with different techniques, synthetic flocculation is viewed as a standout amongst other strategies for gathering cells, since it can deal with a lot of microalgae, can be utilized with a wide assortment of animal varieties, is dependable and reasonable.

Expenses of inorganic flocculants are much lower than those of natural. In any case, bigger amounts required by inorganic flocculants may include high cost per unit of high natural flocculant microalgae. Sukenik et al. revealed that the ideal flocculant dosage required for inorganic ocean growth microalga flocculates was 5-10 times higher than that required to coast the freshwater microalgae. Shamma likewise noticed that the higher cost of natural coagulant can be repaid with low measurements required than inorganic flocculants.

Microalgae gathering methods utilizing compound flocculation are not environmental in light of the fact that they bring chemicals into the framework by expanding broke up solids and evolving shading. Inorganic flocculants can be harmful and may effectsly affect the microalga by adjusting its development media and changing its shading, avoiding water reuse and reusing. Distinctive analysts expressed that flocculant chemicals that are poisonous and cancer-causing and are not reasonable for collecting biomass microalgae in handling for nourishment supplements and sustenance added substances. Accordingly, the flocculant choice ought to be founded on cost, harmfulness and media reuse.

Auto-flocculation Harvesting Methods

A few types of microalgae can flocculate immediately, as a known procedure, in light of certain natural burdens. This marvel is known as autofloculation.

Types of Environmental Stress:

There are a few factors that influence the effectiveness of autoflocculation, which include:

- pH,
- Dissolved oxygen content
- Nitrogen fixation
- Amount of calcium and magnesium particles in arrangement.

PH

At the point when the pH of the medium builds, the cells are pooled and settled by gravitational power. The expansion of more bases in the medium expanded the arrangement of thick chips bringing about circumstances not as much as decantation. Notwithstanding, it is vital to take note of that not all flocculation species with high pH levels.

Seen With a pH of 10.5, 90% flocculation proficiency was resolved for freshwater species Chlorella vulgaris, Scenedesmus sp. E Chlorococcum sp. what's more, pH 9.0-9.3 ha Determined 90% flocculation productivity for marine species Nannochloropsis sp. E Phaeodactylum.

Broken down Oxygen

It has been that in some flocculation types of microalgae can normally happen with varieties in broke down oxygen focus. It has been accounted for that the broke down oxygen stress can prompt flocculation of microalgae. Increment in oxygen broke down in arrangement triggers auto-floculation by expanding the coupling locales accessible on the cell surface. Significant restricting destinations cause monstrous cell arrangement that expands the heaviness of the staple and builds the sedimentation rate. They likewise noticed that expanding the photosynthetic movement of microorganisms builds the broke up oxygen content and the development of thick pieces. Wilen and Balmer watched that substantial flocculates can be produced when the broke down oxygen fixation is high in the medium. A scientist noticed that the disintegrated oxygen centralizations of 14-16 mg/l flocculation advancement in the framework.

Nitrogen

Programmed flocculant microalgae cells can be enacted, obviously, because of ecological anxiety caused by the convergence of nitrogen. Shelef Sukenik and found that a few types of microalga glide each other because of nitrogen worry in the media. Becker watched that microalgae cells can be included because of nitrate osmosis. Osmosis of nitric nitrogen expands the pH of the medium and advances cell flocculation. Nurdogen and Oswald additionally watched that the osmosis of nitrates caused auto-flocculation of microalgae.

Expansion of Ca2+ and Mg2+

Programmed floculation happens suddenly in the way of life media coming about because of the coprecipitation of calcium and magnesium salts introduce in the medium bringing about the pH variety of the medium. Smith and Davis assessed Mg2 +, Ca2 + and CO32-particle by their flocculation and sedimentation impact of microalgae cells and found that the Mg2 + ionic with abnormal amounts of pH brought effective flocculation and fast sedimentation. Settlement rates have been 100 times higher than those acquired by sedimentation. The purpose behind this wonder is that magnesium hydroxide chips are stacked decidedly while floconate calcium carbonate is stacked adversely. In this way, destabilization of the cells is adversely charged higher microalgae utilizing magnesium rather than calcium. The ideal pH for autofloculation is a particular strain. Nguyen et al. he revealed that the Chlorella vulgaris autofloculó species with a productivity of 90% by including Ca2 + and Mg2 + at centralizations of 120 mg/l and 1000 mg/l, separately. Vandamme et al. he noticed that the expansion of Mg2+ in the way of life of vulgaris Chlorella auto-floculación incited.

Favorable circumstances and Disadvantages

The utilization of pH flocculation is favorable by the way that flocculated cells initiated by pH were indistinguishable to non-flocculated microalgae cells. This implies this procedure has autofloculación low shear quality in the cells when contrasted and centrifugation. Knuckey et al. discovered that chlorophyll a T. pseudonana culture was in place utilizing prompted pH techniques, yet the microalgae centrifuged cells had just mellow chlorophyll tops. This is essential when microalgamic biomass is required for use as an eating regimen count calories or to utilize the extraction of specific mixes, (for example, chlorophyll an) of the cell. They likewise found that the types of microalgae taken (utilized for shellfish sustain) utilizing pH-actuated flocculation are a superior decision of eating regimen contrasted with harvests by centrifugation and much better flocculation by ferric chloride. It has likewise been accounted for that the pH-actuated development of C. muelleri to sustain P. mondon shrimp was just marginally slower advancement rate than those utilizing crisp C. muelleri.

Bio-Flocculation Harvesting Methods

He considered the utilization of microorganisms for the recuperation of microalgae biomass (Table 6.3). This technique works by adding microorganisms to holding fast to microalgae culture cells, causing weight pick up and prompting cell creation at the base of the vessel. The supernatant containing the way of life medium was tapped and washed with water to decrease saltiness.

Table 6.3: Use of differentfungi and bacteria microorganisms for the bio-flocculation of microalgae

Microoganism	Bio-Flocculated Microalgae
Bacteria	
Bacillus licheniformis	*Desmodesmus sp.*
P. stutzeri&B.cereus	*Pleurochrysiscarterae*
Paenibacillussp.	*Chlorellavulgaris*
Paenibacilluspolymixa	*Scenedesmussp.*
Bacillus subtills	*Chlorellavulgaris*
Fungi	
Ankistr odesmusfalcatus	*Chlorellavulgaris*
Scenedes musobliquus	*Chlorellavulgaris*
Tetraselmissuecica	*Nannochloropsisoleabundans*
Skeletonema	*Nannochloropsis*

This is the viable flocculation of Chlorella utilizing bio-flocculant bacterial species. Clorellavulgaris effectively gathered utilizing the woolly bio-bacterium Paenibacillus sp. Kim et al. Successful flocculation of the species Scenedesmus sp. Utilizing bio-flocculant Paenibacillus polymyxa. Ndikubwimana et al. have collected the types of microalgae Desmodesmus sp. Bacillus licheniformis microscopic organisms utilizing 98% evacuation effectiveness. Zhang Hu expressed that developed Chlorella vulgarisspecies with different filamentous parasites and separated microbial transesterification oil biodiesel.

Variables Affecting Bio-flocculation

The components influencing bio-flocculation incorporate convergence of the bio-hairy, pH and the selectivity of the microorganism.

Bio-hairy Concentration

The bio-flocculation administration relies upon the proportion amongst microalgae and non-flocculant bio-flocculation species. Bioflocultivos focuses that are more noteworthy than the convergence of non-woolly microalgae increment the sedimentation rate. Lee et al. They found that the expansion of woolly microalga culture microscopic organisms did not build sedimentation rate. Salim et al. he gathers non-flocculant microalgae effectively by including biofloculadoras species and found that the expansion of microalgabiofloculadoras sedimentation of microalgae and expanded harvest proficiency incited. Ohet al. They announced that the effectiveness of flocculation of microorganisms Paenibasillus vulgarisusing C. sp. It reduced with expanding bacterial weakenings. Zheng et al. he announced that the adequacy of flocculation of C. vulgaris utilizing B. Subtilis bio-flocculant expanded with expanding centralizations of C. vulgaris biomass. It has been accounted for that C. vulgaris flocculation proficiency expanded as the centralization of microbes (Flavobacterium, Terrimonas and Sphingobacterium) expanded in culture.

PH: It has been watched that the bioflocculation proficiency is affected by the pH of the dirt. PH adjusts the surface charge of the atoms of the medium decides the level of fascination/repugnance. Resaerchers revealed that the proficiency of flocculation of C. vulgaris utilizing microscopic organisms Paenibasillus sp. Expanded with pH increment from 5 to 11. Ndikubwimana et al. he noticed that the flocculation effectiveness of Desmodesmus sp. It has developed from 43 to 98% utilizing microscopic organisms of Bacillus licheniformis when the pH diminished from 7.2 to 3. Lee et al. It has been watched that immaculate C. vulgaris societies demonstrated no flocculation with the pH increment of

3 to 11, yet with bacterial societies indicated more noteworthy flocculant proficiency (43 to 94%), with pH increment in the 3-11. Be that as it may, Zheng et al. noticed that the bio-flocculation productivity with B. subtilis with microalgae Chlorella vulgarisprotothecoides species Chlorella and was seen by pH.

Species Selectivity: The bacterium Paenibasillus brought about a flocculating proficiency of 83% and 95% with Chlorella vulgarisand Scenedesmus sp., separately.

Favorable circumstances and Disadvantages

The advantages of utilizing bio-flocculants incorporate biodegradability, non-poisonous quality, and intermediates framed amid debasement there are optional toxins. Salim et al. noticed that a two-arrange accumulation process utilizing characteristic microalga flocculants to incite non-flocculant microalgaas took after by centrifugation, decreases vitality utilization by 13.8 MJ kg (dry weight)/min of 2 MJ/kg (dry weight).

The inconveniences of this strategy are that flocculant microorganisms are particular algae and reusing and recuperation of these life forms can be troublesome. Ohet al. He said that bio-flocculants used to dry out microalgae cells ought to be tried for intense oral danger; recuperated biomass is utilized as a part of sustenance added substances and dietary supplements. Vandamme et al. he demonstrated that the utilization of parasites or microscopic organisms as flocculant operators brings about microbiological defilement of the microalgal biomass that should be assessed before use in bolster or nourishment items.

Electrophoresis Harvesting Methods

Electrolytic techniques are utilized to dispose of the utilization of exorbitant and harmful chemicals, for example, microalgae carry on fundamentally the same as colloidal particles, enabling them to isolate from a watery vehicle from the electric field.

Sorts of Electrophoresis

The fundamental electrophoresis strategies that can be utilized for collecting microalgae are:

- Electrolytic coagulation

- Electrolytic flocculation
- Electrolytic buoyancy.

Electrolytic Coagulation

This kind of electrophoresis strategy requires the utilization of physical and synthetic boosts for viable detachment of microalgae biomass. The coagulation procedure is actuated by current era from an iron or aluminum terminal, as appeared in Figure 6.14.

The measure of electrical current going through the fluid medium decides various metal particles broke up in the fluid suspension. Metallic particles are discharged into the arrangement hydroxides metal add to the destabilization of the colloidal suspension and coagulate biomass. Flocculation is acquired by restricting negative metal positive negative metal microalgae cells and moving to the anode due to theelectrophoretic movement. Quick coagulation coming about because of high current thickness, however the cost related with the technique is high.

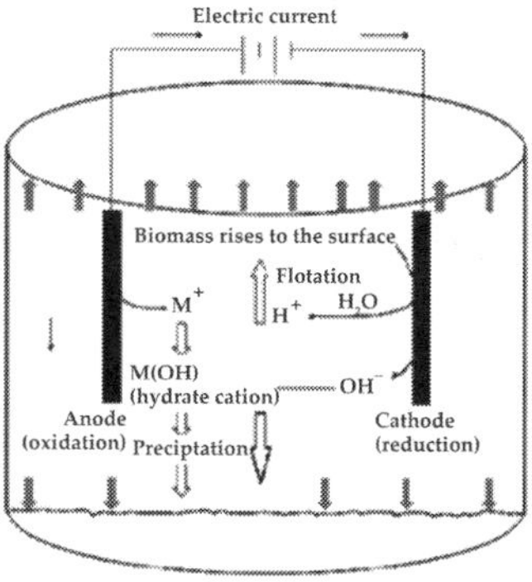

Figure 6.14: Dewatering microalgae through electrolytic coagulation.

SOURCE:https://www.intechopen.com/books/new-technologies-in-the-oil-and-gas-industry/electrolytic-treatment-of-wastewater-in-the-oil-industry

Uduman et al. he utilized a 5 V aluminum cathode set and accomplished efficiencies of 93.3% and 87.3% electrocoagulation at 600 s for the Tetraselmis sp. also, Chloroccum sp., consciously. Azarian et al. Recuperate 99.5% of aggregate suspended solids in 15-minute electrocoagulation utilizing a 550 Watt control source with thealuminum anode. They likewise found that a 100 W control supply required 30 minutes to get comparative outcomes. A 99% Escherichia coli evacuation is accomplished in 20 minutes by electrocoagulation utilizing an aluminum cathode with a 12 W influence supply.

Electrolytic Flocculation

Electrolytic flocculation moves negative cell development towards the anode, as appeared in Figure 6.15. Around then, cells lose their chip shaping charge that can be lifted to the surface holding fast to bubbles framed by electrolysis of water.

Scientists in examine tried electrolytic flocculation effectiveness in a 100-liter holder furnished with vertical terminals and watched 80-95% evacuation proficiency in 35 minutes.They additionally noticed that the rate of expulsion diminishes as declines microalgae pressures and less vitality is expended when the aggregate range of the anodes has been lessened and additionally diminished the separation between the cathodes.

93.6% Botryococcus brani recuperation productivity is acquired following 30 minutes utilizing an electrolytic flocculation procedure with a 6 W. It was watched marine microalgae recuperation at 85% and 95% following a hour utilizing electrolytic flocculation with A power supply of 5 VY 5.2 V individually. Zenouzi et al. 97.4% Dunaliella saline expulsion proficiency was accounted for after 3 min utilizing electrolytic flocculation.

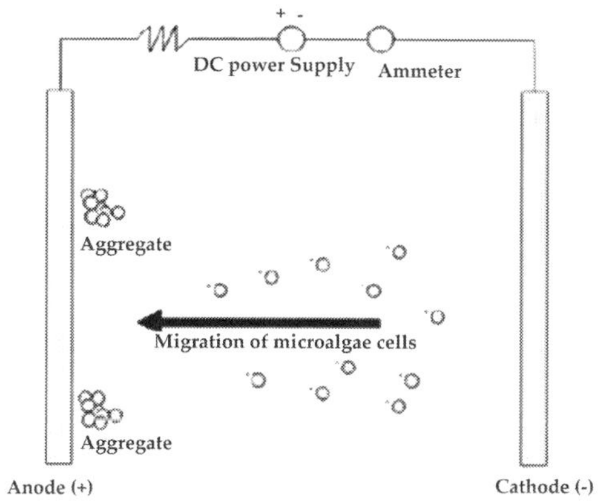

Figure 6.15: Electrolytic flocculation for dewatering microalgae

SOURCE:https://www.intechopen.com/books/new-technologies-in-the-oil-and-gas-industry/electrolytic-treatment-of-wastewater-in-the-oil-industry

Electrolytic Flotation

This strategy is like the dynamic metal anodes coagulation electrolytic used to coast microalgae cells. The distinction between the two methods is that the cathode in the electrolytic buoyancy is made of a latent metal (steel) which is not electrochemically stored as appeared in Figure 6.16. Hydrogen shapes latent metal rises by electrolysis of water. The suspended particles are joined to the gas air pockets and ascend to the surface of the holder.

Research have considered electrolytic microalgae buoyancy cell utilizing an anode amalgam and titanium polyvalent aluminum cathode and found that expanding the electrolysis time diminishing force and expanding the rate of chlorophyll expulsion. It was additionally noticed that the measure of chlorophyll estimation is identified with the convergence of microalgae evacuated by electrolysis. The utilization of high power is restricted by the higher warmth and pH increment.

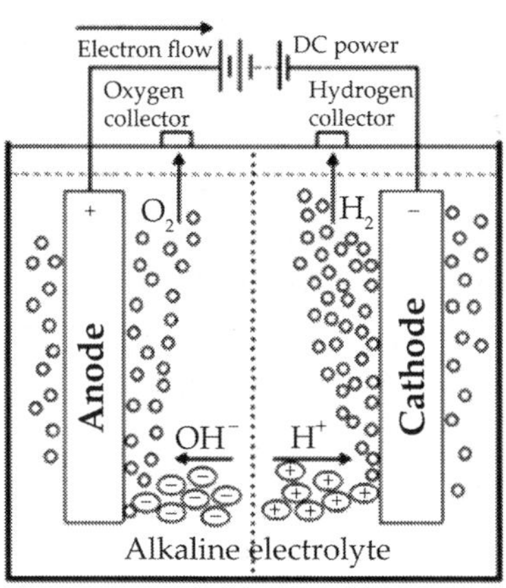

Figure 6.16: Electrolytic flotationfor dewatering microalgae.

SOURCE:https://www.google.com/search?client=opera&hs=apa&biw=1024&bih=636&tbm=isch&sa=1&q=+electrolytic+flocculation&oq=+electrolytic+flocculation&gs_l=psy ab.3...32297.32297.0.32634.1.1.0.0.0.0.0.0..0.0....0...1.1.64.psy-ab..1.0.0. VtJ Quv3yQFs#imgdii=1l2VnKrx9xofKM:&imgrc=MkXguZxnQDUPrM:

Energy Consumption

Lack of hydration of microalgae by electrophoretic systems requires less vitality (0.2 to 2.1 Wh/g) than other accumulation strategies (0.18 to 35.62 Wh/g). It has sbeen detailed that the measure of vitality required for flocculant microalgae C. vulgaris freshwater and marine species P. tricornutum was 2.1 Wh/g 0.2 Wh/g (dry weight), individually. González-Fernández and Ballesteros said marine microalgae require less vitality to gather, as the marine condition takes into account more prominent conductivity that supports the electrocoagulation procedure. Kim et al. He detailed that power utilization trade extremity utilizing two sorts of anodes changes from 1.19 to 1.23 Wh/g for recovery microalga 99% gathering after 15 min. Microalgae drying out by microalgae culture of 0.8-1.5 Wh/g were watched, while extraneous filtration, weight channels, vacuum channels, centrifugation and flocculation with polymers devoured 3.47, 0.18, 1.19, 1.67 And 35.62 Wh/g, separately.

Favorable circumstances and Disadvantages

The advantages of electrophoresis gathering incorporate flexibility, effectiveness, wellbeing, selectivity, ecological similarity, and execution. The base vitality is devoured when the ideal potential contrast (0.331 kWh/m3) is utilized to control the surface of the anode and the separation between the terminals. There are no extra expenses related with flocculants items. Expenses related with drying out with microalgae electrolytic techniques were essentially lower than other gathering strategies, for example, flocculant sedimentation, flocculants centrifugation, and buoyancy. This shows, albeit more noteworthy power devoured by electrolytic strategies, the accumulation cost is much lower than other gathering systems. A portion of the weaknesses related with this gathering system incorporate fouling cathode and mobile phone adjustment. Current reductions by 5-10% after re-utilizing of cathode because of inside resistance and changes in cell sythesis can be actuated utilizing high current thickness.

The mix of Methods

Analysts audited a few accumulation strategies and found that the blend of flocculation sedimentation or filtration or buoyancy turning centrifugation for monetary options. It has been accounted for that vitality can be put away and the cost can be decreased microalgae gathering a two-stage strategy utilizing two procedures. At first, microalgae aggregated at 2-7% of suspended aggregate flocculation solids and cells (suspended solids convergence of 15-25%) was additionally thought to a glue by an optional filtration or electrophoresis gathering step. Funk et al. Buoyancy with broke up air incorporated with substance flocculation (ferric sulfate) and watched expanded recuperation proficiency of 88% to 95% for Chlorella vulgaris.

Kim et al. revealed that another and imaginative innovation to enhance the method of electrophoresis forms is the blend of electrolytic and electrolytic microalgae coagulation constant change electrolytic. In this strategy, the bearing of current (extremity trades) for constant gathering of microalgae crops is traded. The present course makes two stages utilizing a couple of anodes. The principal organize attempts to destabilize negative microalgae cells shaping pieces. Drop development is interceded by metal particles that are discharged by the anode that breaks down in arrangement. In the second stage, the era of blocked metal particles and air pockets shaped by both flocculant cathodes ascend to the start of the arrangement by gliding.

Xu et al. utilized the buoyancy of the scattered incorporated air and watched 98.9% gathering productivity in 14 min electroflocculation. They found that the total cells expanded with the restriction consolidated into those saw with electrofloculation framework. The utilization of streamlined buoyancy rate expanded total arrangement. Be that as it may, the aerating and cooling stress proceeded in the framework exasperates the total increment of Floc of algae. So the aggravation was maintained a strategic distance from by capturing the air reconciliation into the framework once the total size pinnacle.

Comparative Analysis

Selection of Criteria

Eight criteria were utilized for the assessment of microalgae reaping systems:

(a) Dewatering effectiveness

(b) Cost

(c) Toxicity

(d) Suitability for extensive scale utilize

(e) Time

(f) Species specificity

(g) Reusability of media

(h) Maintenance.

These criteria were chosen in light of the data in the writing on these microalga accumulation strategies. Near examination was performed utilizing these criteria to decide the most proficient, financially savvy and environmental drying strategy for an extensive variety of microalgal species that is reasonable for huge scale application?

BIOREMEDIATION

Bioremediation is a procedure of treatment that utilizations characteristic microorganisms (yeasts, parasites or microscopic organisms) to disintegrate or debase risky substances into less harmful or non-lethal substances. Microorganisms, similar to people, eat and process natural substances for supplements and vitality. In substance terms, "natural" mixes are those containing carbon molecules and hydrogen.

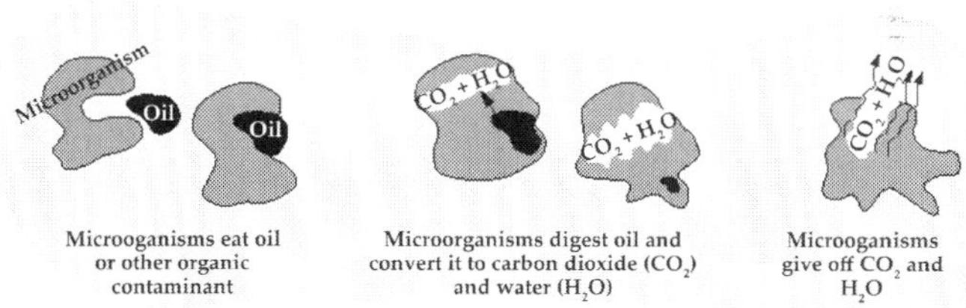

Figure 6.17: Schematic Diagram of Aerobic Biodegradation in Soil

SOURCE:http://staff.icar.cnr.it/spezzano/colombo/bioris/cortesto.htm

A few microorganisms can process natural substances, for example, fills or solvents that are perilous to people. Microorganisms decay natural contaminants in innocuous items, particularly carbon dioxide and water (Figure 6.17). Once the contaminants are corrupted, the number of inhabitants in microorganisms is lessened on the grounds that they have utilized their whole sustenance source. Dead microorganisms or little populaces without sustenance display no danger of defilement.

Bioremediation is Important

Bioremediation is imperative for two reasons.

1. It utilizes no chemicals

An issue with the utilization of manufactured chemicals in the treatment and disposal of contamination is that the chemicals perhaps make in water. There have been numerous chemicals utilized toward the start of the waste administration time frame that we now know is extremely harming to plants and creatures and human life, once got in the water organize.

2. It can enable waste to be reused:

Another real reason that bioremediation is favored is that once the waste is dealt with and the sullying killed or evacuated, the waste itself would then be able to be reused. At the point when concoction remediation sorts are utilized, the waste is as yet sullied just with a less dangerous substance and as a rule, can't then go into the reusing procedure. Bioremediation takes into account more waste to be reused while compound strategies still make squander that can't be utilized and must be put away some place.

Classes of Bioremediation utilized

There are two classes of bioremediation utilized. Try not to confound the class sort with the real sorts of bioremediation accessible, the classes portray the general use of the life forms.

The two classes are:

- In-situ
- Ex-situ

In-situ

In situ alludes to when polluted waste is dealt with comfortable purpose of cause. For instance, there might be soil that is polluted. Instead of expel the dirt from its purpose of birthplace, it is dealt with right where it is. The advantage to in situ treatment is that it keeps the spread of pollution amid the dislodging and transport of the sullied material.

Ex-situ

Ex situ alludes to treatment that happens after the tainted waste has been evacuated to a treatment zone. To utilize soil as the illustration once more, the dirt might be evacuated and transported to a range where the bioremediation might be connected. The principle preferred standpoint to this is it contains and control the bioremediation items, and also making the range that was defiled accessible for utilize.

Sorts of Bioremediation

There are much more than 9 sorts of bioremediation, however the accompanying are the most widely recognized courses in which it is utilized.

1. Phytoremediation

Utilize the plants to evacuate contaminants. The plants can draw the contaminants into their structures and clutch them, successfully expelling them from soil or water.

2. Bioventing

The dirt is utilized to build oxygen rates in the waste. This is a powerful approach to kill certain oxygen delicate metals or chemicals.

3. Bioleaching

Living beings used to expel metals from soil. Certain sorts of life forms are drawntoo overwhelming metals and different contaminants and assimilate them. One new approach was found when angle bones were found to pull in and hold overwhelming metals, for example, lead and cadmium.

4. Landfarming

Turning tainted soil for air circulation and filtering to expel contaminants, or intentionally draining a dirt of nitrogen to evacuate nitrogen based life forms.

5. Bioreactor:

The utilization of uniquely outlined compartments to hold the waste while bioremediation happens

6. Fertilizing the soil

Characteristic rot and remediation process containing waste

7. Bioaugmentation

Adding microorganisms and life forms to fortify the same in waste to enable them to assume control and clean the region

8. Rhizofiltration

In plants, water is utilized to expel metals

9. Biostimulation

The utilization of organisms intended to evacuate sullying connected in a medium to the waste.

Factors Responsible for Microbial Bioremediation to be Effective

The significant favorable position of the bioremediation strategies is that it takes into consideration defilement to be dealt with, killed or evacuated and afterward delivers a waste item itself that is all the more effectively discarded. Sometimes, there is no requirement for transfer by any stretch of the imagination. On account of the plants utilized as a part of phytoremediation and rhizofiltration, the plant can accomplish something many refer to as bioaccumulation. This implies it clutches the contaminant. As the plant is as yet developing, there is no compelling reason to expel and pulverize it. From numerous points of view, it is like having a rechargeable battery. On account of sullied squander, the plant continues developing to take into consideration more stockpiling of waste. This is an interestingly practical answer for polluted waste.

- **Microbial Population**: Suitable sorts of living beings that can biodegrade the majority of the contaminants
- **Oxygen**: Enough to help high-impact biodegradation (around 2% oxygen in the gas stage or 0.4 mg/liter in the dirt water)
- **Nutrients**: Nitrogen, phosphorus, sulfur, and different supplements to help great microbial development
- **Water:** Soil dampness ought to be from 50– 70% of the water holding limit of the dirt
- **pH**: Best range is from 6.5 to 7.5
- **Temperature**: Appropriate temperatures for microbial development (0– 40°C)

Sullying

There are a few sorts of sullying that are exceptionally hard to utilize bioremediation for. The two greatest concerns are:

1. Cadmium
2. Lead

Both of these are delegated overwhelming metals and are hard to expel utilizing microorganisms. As specified before, a current disclosure about the retention rate of fish bone needs to demonstrate fruitful. Truth be told; bone appears to hold the piece of information for evacuating substantial metal tainting. Burn is accustomed to evacuating little measures of zinc, lead, and cadmium; and it is believed that the calcium in the fish bone is the thing that makes it compelling.

Utilize Bioremediation on Nuclear Waste

1. **Bioreactor**: Nuclear waste is now contained inside vessels that keep the sullying from spreading. This turns into a bioremediation as a result of what at that point occurs inside that compartment.
2. **Composting**: When you consider it, atomic waste experiences precisely the same as thematerial you put in a treating the soil heap. The waste has all that it needs to separate itself, it just

takes any longer. When it is secured in the bioreactor vessel, the normally happening procedure of bio remediation assumes control.

Wellspring of Bioremediation

Some portion of the preservation endeavors by researchers are concentrating on the compelling utilization of bioremediation on polluted waste. Defiled waste originates from two distinct causes:

1. Natural
2. Man-made

There are various characteristic sullying forms that happen as a material or substance separates. As it travels through its phases of deterioration, fluctuating levels of poisons might be available. That man is fit for making a gigantic measure of sullying is outstanding. However remediating them with unnatural substances has ended up being the arrangement. At the point when squander is bioremediated it can be reused, or if nothing else, put away such that it makes no further damage theenvironment around it. Masters are always looking towards the common rot and remediation procedures to see how they could be controlled to help oversee man-made waste. This will help keep promote contamination from squander, or the items that treat it. Bioremediation is a standout amongst the most financially savvy and safe arrangements we have now to oversee sullied squander.

Examples of Bioremediation

Of course, you most likely found out about bioremediation in your secondary school science class. However, how can it apply to certifiable circumstances? There are a few branches of bioremediation, each with its own specific strategies and capabilities. It is critical to take note of that while these branches share a similar title, bioremediation, they are taken care of distinctively and their administrations don't cover.

1. **Crime scene cleanup**: Bioremediation in this sense includes the cleanup of blood and organic liquids that can act wellbeing dangers such like hepatitis, HIV, and MRSA. As opposed to utilizing standard cleaning specialists like dye or smelling salts, wrongdoing scene cleaners utilize chemical cleaners to

free the scene of hurtful substances. The organization that has some expertise around there of bioremediation and has very nearly 20 years of involvement in the field. Fallout does not remediate natural contaminations.

2. **The cleanup of debasedsoil:** Human movement has brought numerous poisonous substances into nature's dirt and groundwater. As indicated by a paper distributed by Montana State University, "Amid bioremediation, organisms use compound contaminants in the dirt as a vitality source and, through oxidation-diminishment responses, process the objective contaminant into useable vitality for microorganisms."
3. **Oil spill cleanup**: You may recall the Deepwater Horizon oil slick that occurred in 2010, where 3.19 million barrels of oil spilled in the Gulf of Mexico. Because of the adequacy and lower cost of bioremediation, two techniques were utilized to tidy up after the Deepwater Horizon oil slick.
   - **Bioaugmentation.** The infusion of a little measure of oil-corrupting organisms into an influenced region.
   - **Biostimulation**. The expansion of supplements to animate the development of intrinsic oil-debasing organisms to expand the rate of remediation.

CHAPTER 7

METHODOLOGY FOR COST ANALYSIS OF MICROALGAE PRODUCTION

Model systems for the growth of engineering systems and the collection of ATS (Algal Turf Platform) and ORP (Open Race Pond) were generated. Biomass production for each system was comparable, 500 kg of dry weight ash per year. The modeling and results of each of the production systems have been divided into two efforts to assess the costs associated with biomass production and the extension of labor to assess fuel production within the boundaries of the defined system as 1) Figure 7.1. The first limit, the growth system, limited to the production and the collection of biomass at 20% of solids. The second limit, the biofuel system, extends the work including the production of renewable diesel through HTL and hydraulic processing. Energy flows and process modeling mass engineering combined with economic modeling to evaluate the feasibility of production through alternative paths and compare the growth of two architectures. The results are presented at a cost per tonne (t) of 20% of solid dry ash without biomass for the first limit of the system and the cost per gallon of renewable diesel for the second system boundary with all the results presented in dollars 2014.

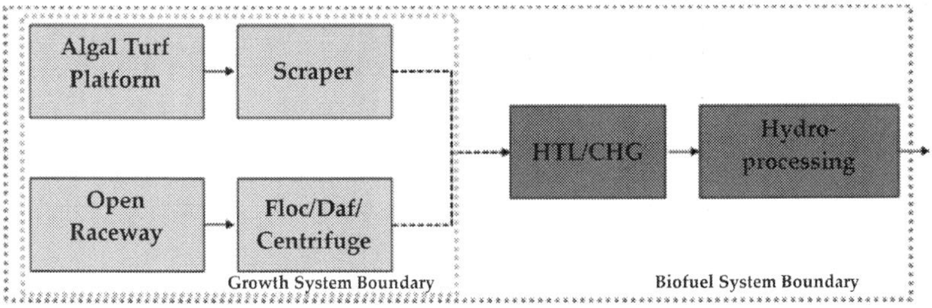

Figure 7.1 System boundaries allowing for comparisons of ATS and ORP growth systems on the metrics of biomass cost per tonne and fuel cost per gallon.

ATS Growth System

System modeling has been used to develop and evaluate the growth of microalgae using an ATS system. The main inputs for the growth of ATS system are listed in Table 7.1. The growth rate was set at 20 g m-2 d-1 based on data provided by the Sandia Laboratories National (BN). ATS module length was fixed at 152 m (500 ft), with each ATS unit set to 405 hectares (1000 acres). In the additional material, a schematic drawing of the system is presented. ATS systems are designed to integrate with water contaminated systems and do not require additional nutrients to add to the system. It is assumed that the system designed limits the water delivery in a single passage. The required hydraulic load is 124 lpm per linear meter. It was assumed that pumping efficiency was 67% with an elevation height of 4 m. The required power is calculated with an equation where g is gravity, RH is the hydraulic load rate, n is the number of ATS modules, Wm is the width of each module and η is the pumping efficiency:

$$P_{ph} = \frac{g * R_H * n * W_m}{\eta}$$

The resulting power is 0.80MW m-1. Accounting for 14 hours anoverall pumping day results in an energy consumption of 4100 MWh m-1 y-1. The ATS system is designed with a gradient of 0.5% and uses a 3D lining fixing substrate and the screen to provide a structure for improving productivity. Capital costs associated with excavation works, roads, pipelines, and land are \$ 371K, \$ 297K, \$ 863k, and \$ 3M for

405 hectares respectively. Screening and screened cost costs represent a significant capital investment of $ 22 million for 405 hectares. Costs associated with a load of nutrients due to the assumption that ATS uses contaminated waste water systems and to provide necessary nitrogen, phosphorus and carbon. Detailed capital assumptions are included in the supplemental material.

Table 7.1: Baseline assumptions for the ATS and ORP growth systems for annual production of 500 ktonne of ash free dry weight

Inputs	ATS	ORP	Description
Algae Growth Rate	20	20	$g\ m^{-2} d^{-1}$
Algae Lipid Content	10	30	wt %
Harvest Cell Density	200	0.5	$g\ L^{-1}$
Module Length	152	100	m
Slope	0.5	0	%
Pond Depth	-	20	cm
Evaporation Rate	-	0.4	$cm\ day^{-1}$
Amount of water recycled	-	90	%
CO_2 Losses	-	10	%
Total Harvesting Efficiency	90	90	%
Ash Content	50	8	%
Nutrient Requirements:			
N, dry wt%	-	7.6	wt %
P, dry wt%	-	0.8	wt %
CO_2	-	1.93	$kg\ biomass^{-1}$

ATS Harvest

An advantage of the ATS system is the simplicity of the biomass collection. The crop is obtained using current farming operations using a tractor plow with an additional dehydration is not necessary. Collection costs are capital purchasing machines and operating costs based on hourly labor costs and diesel fuel costs. Assumptions included required by the ATS 405 hectare module cost $ 350K each machine. It was assumed that the catch rate was 0.84 hectares per hectare-1 with a yield rate of 90% harvest. Operating costs include a diesel company, an alleged cost of $ 3.90 per gallon. The number of employees was about 250% occupational information at the level of trade estimates and specified in the additional material.

ORP GROWTH SYSTEM

Open pit lances have been evaluated by building a system engineering model. Key input hypotheses for system redox growth are listed in Table

7.1. It is assumed that ORP produced at an average growth rate of 20 g m^{-2} d^{-1}. The Redox created using a 4-hectare pool at a depth of 20 cm with a sufficient number of ponds to meet the annual tonnage requirement. The algae circulate through the ponds at a density of 0.5 g L^{-1} using impellers. The power consumption of the impeller is set to 1 kW ha^{-1} for continuous water circulation. The nutrient load is reached by providing nitrogen and phosphorus at 7.6% and 0.8% by weight, respectively. It is assumed that CO_2 is fed through a near power or another source of combustion gas. CO_2 is transferred from ponds to a well with a deflector system to limit degasation. CO_2 sinks charge \$ 8.3k per hectare cost and CO_2 deliveries were set at 250 kWh t-1. It is assumed that evaporation ponds occur at a rate of 0.4 centimeters per day -1. Infrastructure costs were largely collected by Lundquist *et al.* (2010) and Benemann and Oswald (1996) cost 2014 converted into US dollars. It is assumed that the ground needed is low-value land that is not suitable for traditional land crops. Like the ATS system, a \$ 7.4k per hectare was used. The cost for ponds and bladed wheels was set at \$ 34k per hectare. Appliance charges are generally \$ 1k per hectare. The number of employees and their salaries was associated based on a Humbird *et al.* (2011) for a similar plant. Details of personnel costs are included in the supplementary material with a detailed Capital Assumptions table.

ORP Harvest Harvesting

Redox algae are made at a rate equal to the growth rate achieved by the system. A key disadvantage of the redox system compared to the ATS system is the need to dehydrate algae less than 0.1% solids to 20% solids. A three-phase dehydration process is required to provide appropriately dehydrated feed material for the crop. Initial dehydration takes place through sedimentation tanks to obtain a density of 10 g L^{-1} or 1% solids. The suspension is flocculated with chitosan and harvested by a DAF unit to bring the density of 100 g L-1 or 10% solids. Chitosan is required at a concentration of 40 mg L^{-1}. After a centrifuge is used to obtain the desired density is 200 g L-1 or 20% solids for the crop. Capital and operating costs associated with each step crop are based on Wang *et al.* (2007). Harvested operation requirements include lpm 0.078 KW-1 for the DAF unit and 0.066 KW-1 lpm for the centrifuge.

HYDROTHERMAL LIQUEFACTION BIOCRUDE

To study and compare the overall economic viability of growth platforms, the system model has been extended to ATS platforms and ORP growth to include a biorefinery based on the conversion of biomass into renewable diesel through HTL and hydro-thermal as shown from the biofuel limiting system of Figure 7.1.

Table 7.2: Key assumptions associated with fuel production via HTL processing for both the ATS and ORP systems.

Inputs	ATS	ORP	Description
HTL			
HTL Oil Yield	44	44	%
Ash Content	50	8	%
NG Energy	6.2	3.7	M-MJ d^{-1}
Electrical Energy	192	120	MWh d^{-1}
Capital Cost	270	183	M$
Hydrotreating			
Hydrotreating Oil Yield	78	78	%
Processing Capacity	153	153	kgal d^{-1}
Diesel Yield	83	83	%
Naphtha Yield	17	17	%

The results of Jones *et al.* (2014) were used to provide an approximate estimate of operating and capital costs to facilitate comparison with previous works based on a dollar per metric gallon. The ash content of the ATS system is recorded by multiplying a cost factor equal to the amount of mass flow mass needed. Inputs and hypotheses for HTL and the hydrotreatment process are reported in Table 7.2. Renewable hydrotreating diesel fuel has a 78% yield, with 83% being diesel and the remaining 13% of diesel. Catalytic hydrothermal gasification (CHG) is also included to eliminate the carbon content of the aqueous phase after HTL. The treated gas is then used to generate hydrogen in a hydrogen plant that should provide the hydrotreating site. In order to take into account the benefits of each plant, the content of oil and ash has been changed for each system. ATS and ORP algae performance is based on experimental data from Sandia National Laboratories and Pacific Northwest National Laboratory, respectively.

ATS AND ORP BASELINE BIOMASS PRODUCTION COSTS

Capital costs and the operation of ATS and ORP systems were used to assess the economic feasibility of each biomass production system. Costs are presented in Table 7.3. A total cost of $ 510 tons -1 for the ATS system has been determined. Energy costs associated with contaminated water replenishment needed to contribute 48% of total operating costs and are the largest contributor in terms of operating costs. Total operating costs for the ATS system are favorable, as there is no need to dry biomass. The primary costs associated with the collection are fuel and work that contribute 16% and 11%, respectively, which are the main responsible for power requirements. The dominant factors in capital costs include construction costs (infrastructure) and coatings attributed 24% and 27%, respectively. The results of ORP cost systems are presented with capital costs and detailed operation in Table 7.3 at the total cost of the biomass of $ 673 tons-1.

Table 7.3: Detailed cost breakdown for the ATS and ORP growth systems.

Costs Factors	ATS ($/tonne)	ORP ($/tonne)
Operation Costs		
Power	$ 56.21	$ 65.26
Nutrients (N,P,C02)	-	$ 104.25
Fuel (ATS), Flocculant (ORP)	$ 19.14	$ 56.36
Labor	$ 13.30	$ 14.69
Maintenance/Insurance	$ 29.26	$ 25.53
Tax	$ 53.88	$ 56.36
Capital Costs		
Earthworks (ATS), Ponds + paddle wheels (ORP)	$ 7.30	$ 61.50
Liners	$ 90.20	$ 14.93
Pump System (ATS), C02 delivery+sumps (ORP)	$ 15.97	$ 10.64
Piping (ATS), Water/Nutrient/Electrical Supply (ORP)	$ 53.48	$ 17.81
Land Costs	$ 5.59	$ 15.95
Engineering/Tech (ATS), Inoculum system (ORP)	$ 42.70	$ 22.26
Building Costs	$ 7.76	$ 81.27
Other Costs	$ 14.47	$ 30.48
Harvest Systems	$ 81.09	$ 50.20
Site Development	$ 20.27	$ 46.17
Total	**$ 510.65**	**$ 673.65**

Operating costs are dominated by energy requirements, nutrients, and dehydration. The energy needs are driven by the need for continuously circulated ponds and the drainage requirements to obtain the acceptable density of the crops. This translates into energy consumption, which attributes 32% of the operating costs of circulation and dehydration, contributing respectively 85% and 15% respectively. It is also involved in flocculant dehydration costs required for dislocated air flotation that provides 21% of total operating costs. The key factors contributing to the cost of capital include watercourses and pallets, coatings, construction costs (infrastructure development) and site development, accounting for 52% of combined biomass costs. The main differences between the growth of the ATS and ORP system make a difference of 28% of the costs. The ATS system has large initial costs due to ground movements to create a sufficient slope and significant costs associated with a suitable coating system. The capital costs for the system are fueled by ORP needed for the development of ponds, comparable requirements with the ATS system and drainage infrastructure systems. The capital costs for the growth system of \$ 339 tons ATS^{-1} is very comparable to the ORP growth system for a total of \$ 351 tons^{-1}.

The Redox growth system has the disadvantage of being more expensive to use due to dehydration processes to obtain the desired crop and nutrient density. The operating costs associated with the ATS growth system are \$ 118 tons^{-1} while operating costs for redox system growth represent \$ 266 tons^{-1}.The ATS system has an advantage over the ORP system that does not require raw nutrients and uses a growth system bound to improve dehydration of seaweed costs. These benefits make the total cost of biomass significantly lower for the ATS system. The basic case for the redox system includes inputs for the acquisition of minimum CO_2 which is a nutrient necessary for growth and accelerated distribution is limited to the site. Previous resource evaluations have highlighted co-location as the scalability limitation of the factor of biofuel algae systems. Compared to current literature, the ATS system has very promising results at \$ 510 per ton. The \$ 673 tons^{-1} for the resulting redox system is 42% higher than the \$ 474 ton^{-1} assumed by Davis *et al.* (2014a) and Jones *et al.* (2014) with important costs needed for this course. However, the results are in line with the optimized cost of \$ 740 tons^{-1} reported by Norsker *et al.*, 2011. A fundamental prerequisite for performance of systems comparing average annual productivity, taken as 20 gm^{-2} D^{-1} to ATS and ORP. The ATS system is based on dynamic indigenous cultures that have proven to be robust. It has been shown that ORPs are susceptible to invasive species, thus reducing overall system productivity and adversely affecting the economy. When considered to strictly limit the growth system, it is clear that the ATS system has many advantages that lead to a lower overall cost of biomass.

BIOFUEL PRODUCTION COSTS

The system boundary extended to the downstream biomass through HTL. These data confirm the economic evaluation of production platforms in a system of limits covering fuel production. Figure 7.2 shows the costs that arise for ATS and Redox systems.

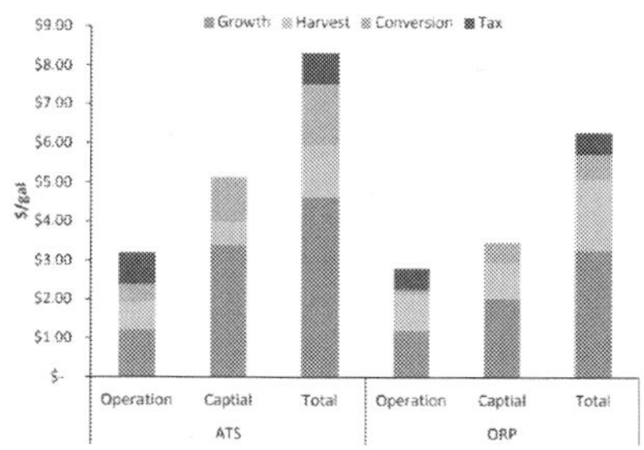

Figure 7.2: Fuel costs for the ATS and ORP systems based on growth, harvest and conversion costs.

When compared with higher yields of lipids and reduced ash content achieved by a redox system. By recognizing these key benefits for the redox system, it is important to realize that due to the robust growth growth bound for the ATS system, ORP is susceptible to invasive species that may cause the system to stop. The costs for both systems illustrate the need for more research and development to produce the product competitively priced. Fuel costs resulting from this study are at the lower end of the results reported in the literature. Fuel costs have been reported starting at $ 2.20 per gallon in Nagarajan *et al.*, 2013 and as high as $ 31.36 per gallon by Richardson *et al.*, 2012.

Fuel costs resulting from this study are at the lower end of the results reported in the literature. Fuel costs have been reported starting at $ 2.20 per gallon in Nagarajan *et al.*, 2013 and as high as $ 31.36 per gallon by Richardson *et al.*, 2012. The ATS $ 8.34 system results to fall in mid-range gallon $ 6.27 per gallon Redox system is at the lower end of fuel cost. The high ash content and lipid associated with the ATS system causes the redox system to have beneficial fuel costs but the cost of biomass is greater. Reducing ash could be obtained with relatively simple processes that lead to reduced fuel costs. A 5% reduction in ash content leads to a total cost of $ 7.88 per gallon, down 5.5% of total costs. The results show that the growth system dominates the total cost $ - $ 1.00 $ 2.00 $ 3.00 $ 4.00 $ 5.00 $ 6.00 $ 7.00 $ 8.00 $ 9.00 Total Captial Operation ATS ORP $ / gal tax conversion crop growth associated with fuel production corresponding to 56% and 52% ATS and redox systems respectively.

SENSITIVITY

A sensitivity analysis for each system was carried out to identify the main factors contributing cost models. The graph for ATS redox systems is shown in Figure 7.3 and Figure 7.4, respectively. Clearly the dominant factors for both systems are lipid content and growth rate. Optimizing these parameters, small changes can greatly influence the economic feasibility of using algae as a raw material for renewable fuels. Reducing costs associated with changes in operating and capital costs can help reduce costs but are not as significant as the results with more efficient and productive algae growth. Exploring individual systems can be observed key factors in both systems. The ATS system is driven largely by coating costs, while the redox system is sensitive to the cost of pond development and blade wheels. Downstream processing through HTL can also have a significant impact on the economy of each system. Using sensitivity, it analyzes a $ 3 per gallon equivalent (ESL) path found for each system. For these systems become feasible in today's world economy, to achieve several innovations to reduce costs to be competitive with the cost of oil-derived fuels. The biggest barrier for the ATS system is reduced ash content. We are working to determine a rinsing process to reduce the ash content before introducing algae into further processing. Another complication factor that could have a strong impact on ATS systems is the aids to remove nitrogen and phosphorus from contaminated waste water. Due to the uncertain cost-benefits that might be related to these scholarships, it would be difficult to rely on these savings over a long period of time to reduce costs to be more competitive. Using sensitivity, it analyzes a $ 3 per gallon equivalent (ESL) path found for each system. For these systems become feasible in today's world economy, to achieve several innovations to reduce costs to be competitive with the cost of oil-derived fuels. The biggest barrier for the ATS system is reduced ash content. We are working to determine a rinsing process to reduce the ash content before introducing algae into further processing.

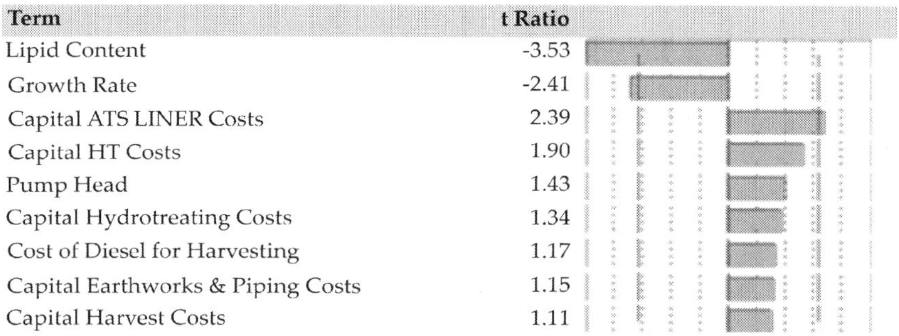

Figure 7.3: Sensitivity analysis results to varying input parameters for the ATS system, t_{crit}=2.36. The top 9 variables are reported.

Another complication factor that could have a strong impact on ATS systems is the aids to remove nitrogen and phosphorus from contaminated waste water. Due to the uncertain cost-benefits that might be related to these scholarships, it would be difficult to rely on these savings over a long period of time to reduce costs to be more competitive. The ATS system could also benefit from better growth rates and lower operating and capital costs. An alternative scenario involving a 74% reduction in ashes content from 50% to 13%, a reduction of 10% of capital costs, 2x subsidy costs for fertilizers and higher productivity, estimated at 30 gm -2 d-1. Savings made by these improvements reduce the base line fuel cost of \$ 7.67 to \$ 3.07 per ESL by GGE. It is expected that some changes in the ATS system will have a greater impact than what can be achieved with the ORP system. This is because the redox system has been developed and optimized for algae growth and many improvements already explored and implemented. The ATS system is not developed as an algae growth system and there is no room for improvement through focused research and development.

Term	t Ratio
Lipid Content	-14.20
Growth Rate	-6.44
Capital HTL Costs	2.64
Capital Ponds & Paddle Wheels	2.38
Capital Harvest Costs	1.32
Nutrient Costs	0.63
Capital Liner Costs	0.58
Cost of CO2	0.42

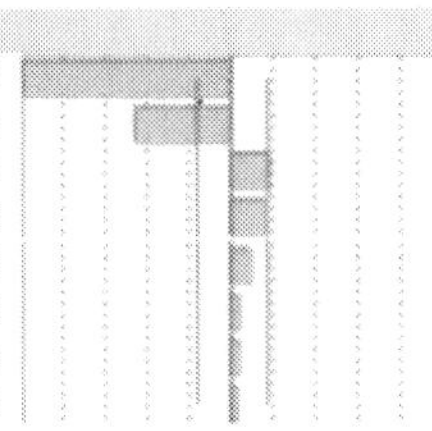

Figure 7.4: Sensitivity analysis results to varying input parameters for the ORP system, t_{crit}=2.36. The top 8 variables are reported.

ORP system for 50% must be executed to reach DOE $3 for ESL. The most important factor to reduce costs would be to improve the rate of growth. There are difficulties associated with the possibility of increasing the rate of growth and ponds are exposed to the weather and can be easily contaminated which leads to culture instability and culture accidents. GMOs or extremophiles are expected to maintain the robustness of culture in these systems. One way to $3 for ESL for Redox includes a 15% reduction in capital and operating costs, a growth rate of 30 g m^{-2} d^{-1}, and a tin system that does not require a coating. Removing a tin coating has been explored and represents a specific geographic limitation. The resulting cost savings to make each improvement cost at $ 3.84 by GG.

CHAPTER 8

ALGAE-TO-BIOFUELS AND PRODUCTS: OPPORTUNITY AND CHALLENGES AHEAD

GREEN FUELS FROM GREEN ALGAE

Feed stocks which are affordable, abundant and sustainable in biofuel industry are important to burgeoning. A significant role can be played by the algae regarding biomass production in lands which are not suitable for agriculture.

Figure 8.1 Algae ponds for green fuel

Source: US DOE/NREL/Gretz, Warren

It can also be important in areas where utilization of unique resources supports a concept of mix fee stocks. Where there is a development for cellulosic biofuel from process engineering lineages direct agriculture, on similar scale there is no development for cultivation of algae. For the improvements and developments regarding algal biofuels, strategic planning & investments are required to support the commercialization of algal biofuels. Uncertainties linked with the large scale production of algal biofuels can be reduced by research and development activities in different areas. In this regards, there are different research and other projects are in process. A detail of some of projects is given below;

Current projects on Algal Biofuels

Department of Energy's primary national laboratory of U.S. is called as NREL (National Renewable Energy Laboratory). This program deals with research & development on energy efficiency and renewable energies' production. A biomass program which deals with the characterization of biomass, thermochemical technologies, biochemical in conjunction with biomass analysis and process engineering. The aim of the project is to produce cost effective, environment friendly and energy efficient which can reduce the dependency on oil, support the rural economy and improve quality of air.

At the Harbor Branch Oceanographic Institution and Woods Hole Oceanographic Institution, algae growth is accelerated by industrial and domestic waste which are rich in organic compounds. Algal cultivation using waste from industries has been explored at University of Georgia, Department of Biological and Agricultural Engineering. A proposal was submitted by the Indianapolis, Indiana to set-up a facility in Cedar Lake. Set-up was proposed to treat the municipal wastewater by algae and production of biofuel from sludge byproduct. A company in Alabama is also using the same approach of algae systems. In San Diego, green crude from algae has been produced by Sapphire Energy. In California, a jet aircraft fuel has been produced from algae.

In Nova Scotia, algae is being cultivated for last fifty years in The Marine Research station inKetch Harbour A 5 million dollar's fund has been provided for this purpose by NRC (National Research Council) in Canada. The purpose of the funding is to develop and build a pilot plant with a capacity of 50,000 L at the facility of Ketch harbor. The station is involved in investigation of number of algae species and assessment of best algae regarding growth in North America regions. For this purpose, Department of Energy U.S and Colorado's National Renewable Energy Laboratory are joined with NRC.

Research in Europe

Following UK Universities are working to produce algae base oil

- University of Sheffield,
- University of Manchester,
- University of Brighton ,
- University of Glasgow,
- University College London,
- University of Cambridge,
- Cranfield University
- Imperial College London
- And Newcastle University.

Microalgae Biotechnology Group, CSIC′s Instituto de Bioquímica Vegetal y Fotosíntesis, Seville is also working on same topic. Regarding algae technologies both in industry and research, the EABA (European Algae Biomass Association) is working with seventy nine members. Headquarter of association is based in Florence, Italy. The main objective of the European Algae Biomass Association is to develop and promote cooperation and mutual interchange for biomass production including utilizations and biofuel uses.

The aim is to develop, create and maintain solidarity and links among the association members to defend their interest at international and European level. The main target is acting like acatalyst to foster synergies between industrialists, scientists and decision makers so as to promote the technology, development of research and industrial capacities regarding algae.

A detailed design study is being carried out by the University of Cambridge and CMCL innovations at Carbon negative Fuels derived from Algal and Solar Technologies (C-FAST) plant. Designing and development of a pilot plant is the main objective, which can help in demonstration of production of fuels based on hydrocarbons (including diesel and gasoline) which can act as raw material and carbon-negative energy carriers for chemical commodity industry. The project reported in June 2013. A biofuel production base don special type of algae was planned by Ukraine.

Algae Cluster Project by the European Commission has been funded under the Seventh Framework Programme, based on 3 algae biofuel projects. Among three, each of project is looking for building and designing a algae biofuel which is different. A land of 10ha is covered by the facility.

Investigation of feasibility for diffrent production procedures (conventional separation/ extraction, gasification, hydrothermal liquefaction and pyrolysis) has been suggested for production of fuel from algae.

Projects in India

A pilot project for production of algal biofuel was commissioned by the Reliance industries and Algenol, U.S in 2014. In India, a protein rich alga which is called as Spirulina is cultivated at commercial scale. In natural/ open oxidation ponds, sewage is treated using algae in India. This provides a biomass which is convertible into biofuel and reduces the BOD of the water.

Research/projects Other Countries

ABO, Algae Biomass Organization which is a not for profit organization. The mission of organization is "to promote the development of viable commercial markets for renewable and sustainable commodities derived from algae". Similarly, an algal researchers' organization (NAA) which is also not for profit, investment community and algae production companies are aimed for the commercialization of algal oil via large scale production of algae.

A forum is provided by National Algae Association, (NAA) to its members for evaluation of different technologies for algae regarding potential early stage company opportunities.

In Ontario, Pond Biofuels Inc., a plant has been setup where algae are produced directly from smokestack from a plant of cement. This biomass is then dried using heat from wheat.

A partnership was announced by the Pond Biofuels with the Canadian Natural Resources and NRC of Canada to build a large-scale bio refinery for algae at a site near Bonnyville, Alberta.

Nova Scotia in Canada has discovered a new algae strain which can produce oil sixty times greater than the other algal biomasses used for biofuel production. This is a big advancement in the field of algal biofuels.

A method for the production of algal lipid in increased quantities has been claimed by the VG Energy. The method works by disruption of metabolic pathways to avoid energy consumption in production of carbohydrates. Company states that lipid production can be increased using such techniques. A nuclear power plant and warm water discharge has been used for the production of algae at Peach Bottom Nuclear Power Station, owned by Exelon Corporation. Genetic engineering has been used for the more effective production of algae by companies like Bio Solar Cells and Sapphire Energy.

According to a research, algae production can be affected in a positive way by genetic engineering. Chemically induced mutations have also been used by Sapphire Energy to produce algae as a crop.

Funding programs for Biofuels

For the promotion of renewable energy, numerous programs have been funded by different agencies. The ecoAgriculture biofuels capital (ecoABC) in Canada provides a grant of 25 million dollars for developing and expending a facility for farmers for biofuel production. An amount of 186 million dollars have been set aside under the project for the future developments in this field.

Similarly, another grant of 500 million Dollars has been applied under sustainable development (SDTC) program for biofuel projects over a period of eight years. Over the last two years, an amount of about 10 million $ has been applied for the research on production of renewable fuels.

Seventh Framework Programme (FP7) is the main funding agency in Europe. The NER 300 which is an independent, unofficial forum has been dedicated fir renewable biofuels. Horizon 2020, is another very important program which will bring research and innovation together for funding of projects o biofuels.

Feedstock Development programof American is dealing with the horizon of production of algae n a sustainable manner.

Nonetheless, with high oil costs and rising attention to the climatic emergency, the two lawmakers, and the overall population, time for the second era of biofuels has never been something more.

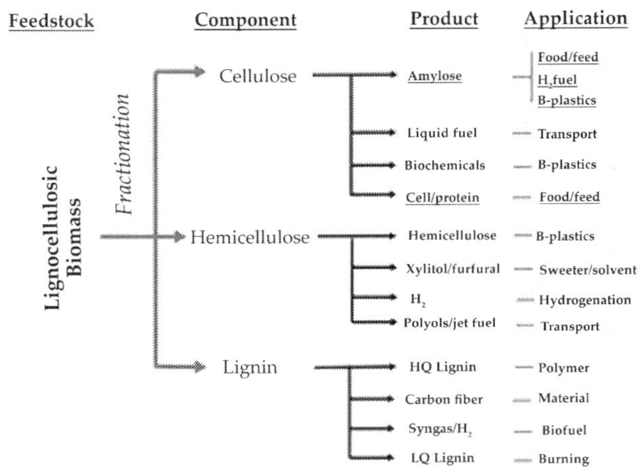

Figure 8.2 Next-generation bio-refineries based on fractionated lignocellulosic components for the production of different products to meet different biochemical and food/feed requirements.

FIVE COMPANIES MAKING FUEL FROM ALGAE

Algenol Biofuels:

The undertaking costs $ 850 million and the organization is devoted to building a green growth cultivate that offers ethanol fuel for $ 3 for every gallon. It is situated at Sonoran Desert (Mexico). The organization is planning to deliver biofuel by utilizing ocean growth without slaughtering or gathering of ocean animals. Creation was started in 2010 and the underlying arrangement was to deliver 1,000 million gallons in initial two years of operation. The generation costs was assessed to be around 85 pennies for each gallon.

Solix Biofuels

Solix biofuel is situated at Coyote Gulch, Colo. The cpmoany was intending to creating a gauge of 3,000 liters of biofuel per section of land every year toward the finish of 2009. Slix biofuels introduced photobioreactors. Under strict control of light and temperature, specific bunches microalgae societies are developed in an expansive shut development council of photobioreactors. These shut frameworks may have a capacity to deliver up to seven times more biomass than open tin frameworks. When societies are completely built up, the oil is removed utilizing synthetic solvents, for example, benzene or ether. The solvents were blended in the rooms to isolate green growth oil and afterward gathered from the surface. Solix is additionally teaming up with the Los Alamos National Laboratory to utilize acoustic innovation to gather algal cells in a thick blend when blown with sound waves. The oil would then be able to be removed from the crushing blend. Usage of this new innovation influences the extraction to process less difficult and conservative by staying away from the requirement for concoction solvents.

Solazyme

Solazyme was built up with an arrangement to supply 400,000 liters of fuel to the Air Force and 190,000 gallons to the US Navy with 1,500 liters of fuel for the US Navy in 2010. The organization is situated in South San Francisco. Solazyme outlines green growth crops utilizing DNA from various strains keeping in mind the end goal to boost the generation of oil. These algal yields develop in measure in the vast aging vessels previously reaping oil. Solazyme claims that thc organization is progressing nicely to deliver more than 20,000 gallons of fuel for the Navy in 2010.

Seambiotic

Seambiotic is 5-hectare business plant and was worked in Israel. Seambiotic develops microalgal societies in open lakes utilizing vent gasses like CO2 and N2 from an adjacent coal plant as feedstocks. Its 1000-square-meter office creates approximately 23,000 grams of green growth every day-three tons of algal biomass would yield around 100 to 200 gallons of biofuel. It as of late framed an association with NASA to improve the development rates of its microalgae. The organization is up and running since 2003. Seambiotic set up a pilot plant in 2005 and a business plant in late 2009.

Sapphire Energy

Sapphire Energy is a 300-section of land coordinated algal biorefinery situated in Southern New Mexico. The organization's attention is on "green rough," a fluid that has an indistinguishable structure from raw petroleum, and is subsequently perfect with existing refineries. The organization has officially demonstrated that its fuel can be utilized as a part of autos and even flies. Sapphire has a 100-section of land pilot office close Las Cruces, N.M. The organization was delivering 1 million gallons of diesel and stream fuel every year by 2011. The future arrangement is to create 100 million by 2018, and 1 billion gallons for every year by 2025.

SCIENTISTS ARE ONE STEP CLOSER TO USING ALGAE AS BIOFUEL

The world is attempting to dispose of petroleum derivatives. It is fundamental that we decrease our carbon impression beyond what many would consider possible in the coming decades, and numerous nations, states, urban areas, organizations, and people focused on that objective. A large portion of the fire has been put on sustainable sources, for example, sun oriented and wind control yet that does nothing for some heritage vehicles and power plants. For them, we require an alternate arrangement: biofuels.

Today most biofuels depend on maize and soybean oil, which is costly to develop and trim. This implies biofuels battle to be focused with customary fills. Researchers have endeavored to locate a decent contrasting option to these products for a considerable length of time, and a group can be found in green growth.

The green growth have various points of interest over maize and soy can develop substantially speedier in all conditions, and is less demanding to gather. The main issue is that the green growth don't create enough fat and oils to make it beneficial as biofuel, so a gathering of specialists utilized hereditary designing to take care of the issue.

Scientists at The Institute for Genomic Research have put in eight years endeavoring to control the genomes of various types of green growth. At a gathering on Monday, Venter himself talked about the task and reported that the group had at long last succeeded. They had dramatically increased the measure of oil the green growth could create.

The hardest piece of the investigation was to recognize the exact quality that controlled the measure of oil delivered by the green growth. The group found the green growth nitrogen to starve, which influenced them to deliver more oil. Through experimentation, the group distinguished the right quality - a controller called ZnCys - around 20 potential hopefuls. At that point, utilizing the CRISPR quality change strategy, the group utilized this quality to compel green growth to deliver more oil. The outcomes were sensational: the normal green growth can deliver around 10 to 15 percent of oil, however the altered variant produces 40 percent.

REFERENCES

1. Azapagic, A., and S.Perdan. 2000. Indicators of sustainable development for industry. A general framework.
2. Process Safety and Environmental Protection 78(4):243-261.
3. An, J.Y., S.J. Sim, J.S. Lee, and B.W. Kim. 2003. Hydrocarbon production from secondarily treated piggery waste-water by the green alga Botryococcus braunii. Journal of Applied Phycology 15(2-3):185-191.
4. AlfaLaval. 2010. Multiple choices for fats and oils refining. Available online at http://local.alfalaval.com/en-us/key-industries/food/fatsandoils/Documents/PFT00355EN_Degumming_neutralization.pdf. Accessed December 21, 2011.
5. Alley, W.M. 2003. Desalination of Ground Water: Earth Science Perspectives. Reston: U.S. Geological Survey.
6. Anderson, A.L., K. O'Brien, and M. Hartwell. 2007. Comparisons of mosquito populations before and after construction of a wetland for water quality improvement in Pitt County, North Carolina, and data-reliant vectorborne disease management. Journal of Environmental Health 69(8):26-33.

7. A.A. Kiss, A.C. Dimian, G. Rothenberg, Solid acid catalysts for biodiesel production towards sustainable energy. Adv. Synth. Catal. **348**, 75–81 (2006)

8. A.A. Kiss, A.C. Dimian, G. Rothenberg, Biodiesel by reactive distillation powered by metal oxides. Energy Fuels **22**, 598–604 (2008)

9. A.A. Kiss, Separative reactors for integrated production of bioethanol and biodiesel. Comput. Chem. Eng. **34**(5), 812–820 (2010)

10. A.A. Kiss, A.C. Dimian, G. Rothenberg, 'Green catalysts' for enhanced biodiesel technology, catalysis of organic reactions. Chem. Ind. Ser. **115**, 405–414 (2006)

11. A.A. Kiss, G. Rothenberg, A.C. Dimian, F. Omota, The heterogeneous advantage: biodiesel by catalytic reactive distillation. Top. Catal. **40**, 141–150 (2006)

12. Assmann, A., A. Braun, S. John, A. Lei, and S. Southard. 2011. The Potential for Micro-Algae and Other "Micro-Crops" to Produce Sustainable Biofuels. Master of Science, University of Michigan, Ann Arbor.

13. Atkinson, K.M. 1972. Birds as transporters of algae. British Phycological Journal 7:319-321.

14. A.K. Singh, S.D. Fernando, Reaction kinetics of soybean oil transesterification using heterogeneous metal oxide catalysts. Chem. Eng. Technol. 30(12), 1–6 (2007)

15. Aresta, M.; Dibenedetto, A; Carone, M.; Colonna, T. and C. Fagale, "Production of Biodiesel from macroalgae by supercritical CO2 Extraction and Thermochemical Liquifaction," Environmental Chemistry Letters, Vol. 3, 2005, pp. 136-139. Briggs, M., "Widescale Biodiesel Production from Algae," Energy Bulletin, October 3, 2004. Brown, L. M. and K. G. Zeiler, " Aquatic Biomass and Carbon Dioxide Trapping," Energy Conversion Management, Vol. 34, No. 9-11, 1993, pp. 1005-1013. Bowman, M.; Hilligoss, D.; Rasmussen, S. and R. Thomas, "Biodiesel: A Renewable and Biodegradable Fuel," Hydrocarbon Processing, February, 2006, pp. 103-106. Aresta, M.; Dibenedetto, A; Carone, M.; Colonna, T. and C. Fagale, "Production of Biodiesel from macroalgae by supercritical CO2 Extraction and Thermochemical Liquifaction,"

Environmental Chemistry Letters, Vol. 3, 2005, pp. 136-139. Briggs, M., "Widescale Biodiesel Production from Algae," Energy Bulletin, October 3, 2004. Brown, L. M. and K. G. Zeiler, " Aquatic Biomass and Carbon Dioxide Trapping," Energy Conversion Management, Vol. 34, No. 9-11, 1993, pp. 1005-1013. Bowman, M.; Hilligoss, D.; Rasmussen, S. and R. Thomas, "Biodiesel: A Renewable and Biodegradable Fuel," Hydrocarbon Processing, February, 2006, pp. 103-106.

16. Adosraku RK, Choi GTY, Constantinou-Kokotos V, Anderson MM, Gibbons WA. 1994.
17. Atmospheric Environment 43:2084-2086. Hossain ABM, Shalleh A, Boyce AN, Naqiuddin M. 2008. Biodiesel fuel production from algae as renewable energy.
18. American Journal of Biochemistry and Biotechnology 4(3):250-254.
19. Armbrust EV. The life of diatoms in the World's oceans. Nature. 2009;459(7244):185–192.
20. Adosraku RK, Choi GTY, Constantinou-Kokotos V, Anderson MM, Gibbons WA. 1994. NMR lipid profiles of cells, tissues, and body fluids: proton NMR analysis of human erythrocyte lipids. Journal of lipid Research 35:1925 – 1931.
21. Atmospheric Environment 43:2084-2086. Hossain ABM, Shalleh A, Boyce AN, Naqiuddin M. 2008. Biodiesel fuel production from algae as renewable energy.
22. Applied Energy 88(11):3632-3635. Karmee SK, Mahesh P, Ravi R, Chadha A. 2004.
23. American Journal of Biochemistry and Biotechnology 4(3):250-254. Karatay SE, Donmez G. 2011.
24. Atmospheric CO2. Mauna Loa. http://co2now.org/images/stories/data/co2-mlo-monthly-noaa-esrl.pdf (accessed on 12th October 2013) NREL. 2012. NREL Creates New Pathways for Producing Biofuels and Acids from Cyanobacteria.
25. Amaro HM, Guedes AC, Malcata FX. Advances and perspectives in using microalgae to produce biodiesel. Appl Energy in press. Doi: 10.1016/ j.apenergy.2010.12.014.

26. Alabi AO, Tampier M, Bibeau E. Microalgae Technogies and Processes for Biofuels/Bioenergy Production in British Columbia. The BC Innovation Council; BC, Canada: 2009.
27. Abou-Shanab RAI, Hwang J-H, Cho Y, Min B, Jeon B-H. Characterization of microalgal species isolated from fresh water bodies as a potential source for biodiesel production. Appl Energy in press. doi:10.1016/ j.apenergy.2011.01.060.
28. Austria: OPEC. Patil V, Tran K, Giselrod HR. 2008. Towards Sustainable Production of Biofuels from Microalgae.
29. Apt KE, Kroth-Pancic PG, Grossman AR. Stable nuclear transformation of the diatom Phaeodactylum tricornutum. Mol Gen Genet. 1996;252(5):572–579.
30. B. M. S. Hossain, A. N. Boyce, A. Salleh, and S. Chandran, "Impacts of alcohol type, ratio and stirring time on the biodiesel production from waste canola oil," African Journal of Agricultural Research, vol. 5, no. 14, pp. 1851–1859, 2010. W. N. N. Wan Omar and N. A. Saidina Amin, "Optimization of heterogeneous biodiesel production from waste cooking palm oil via response surface methodology," Biomass and Bioenergy, vol. 35, no. 3, pp. 1329–1338, 2011. S. Gan, H. K. Ng, P. H. Chan, and F. L. Leong, "Heterogeneous free fatty acids esterification in waste cooking oil using ion-exchange resins," Fuel Processing Technology, vol. 102, pp. 67–72, 2012. E. Lotero, Y. Liu, D. E. Lopez, K. Suwannakarn, D. A. Bruce, and J. G. Goodwin, "Synthesis of biodiesel via acid catalysis," Industrial and Engineering Chemistry Research, vol. 44, no. 14, pp. 5353–5363, 2005.
31. Singh, B.M. Smyth, J.D. Murphy, A biofuel strategy for Ireland with an emphasis on production of biomethane and minimization of land-take. Renew. Sustain. Energy Rev. 14(1), 277–288 (2010)
32. A.F. Turhollow, E.O. Heady, Large-scale ethanol production from corn and grain sorghum and improving conversion technology. Energy Agric. 5(4), 309–316 (1986)
33. B. M. S. Hossain and A. N. Boyce, "Biodiesel production from waste sunflower cooking oil as an environmental recycling process and renewable energy," Bulgarian Journal of Agricultural Science, vol. 15, no. 4, pp. 312–317, 2009.

34. Demirbas, Comparison of transesterification methods for production of biodiesel from vegetable oils and fats. Energy Convers. Manag. 49, 125–130 (2008)
35. Kumari, P. Mahapatra, V. K. Garlapati, and R. Banerjee, "Enzymatic transesterification of Jatropha oil," Biotechnology for Biofuels, vol. 2, article 1, 2009. R. MacEiras, A. Cancela, M. Vega, and M. C. Márquez, "Enzymatic alholysis for biodiesel production from waste cooking oil," in Chemical Engineering Transactions, vol. 19, pp. 103–107.
36. M. S. Hossain, B. A. Nasrulhaq, A. Salleh, and S. Chandran, "Biodiesel production from waste soybean oil biomass as renewable energy and environmental recycled process," African Journal of Biotechnology, vol. 9, no. 27, pp. 4233–4240, 2010.
37. A.K. Agrawal, Biofuels (alcohols and biodiesel) applications as fuels for internal combustion engines. Prog. Energy Combust. Sci. 33, 233–271 (2007)
38. Okullo, A. K. Temu, P. Ogwok, and J. W. Ntalikwa, "The physico-chemical properties of jatropha and castor oils," International Journal of Renewable Energy Research, vol. 2, no. 1, pp. 47–52, 2012.
39. Demirbas, Progress and recent trends in biodiesel fuels. Energy Conserv. Manag. 50, 14–34 (2009)
40. Singh, D. Pant, N.E. Korres, A.S. Nizami, S. Prasad, J.D. Murphy, Key issues in life cycle assessment of ethanol production from lignocellulosic biomass: challenges and perspectives. Bioresour. Technol. 101(13), 5003–5012 (2010)
41. B.M. Choudary, M.L. Kantam, C.V. Reddy, S. Aranganathan, Pl. Santhia, F. Figueras, Mg–Al–O–t-Bu hydrotalcite: a new and efficient heterogeneous catalyst for transesterification. J. Mol. Catal. A Chem. 159, 411–416 (2000)
42. Rice, A. Fröhlich, and R. Leonard, "Bio-diesel production from camelina oil, waste cooking oil and tallow," The Science of Farming and Food, 1998.
43. K. Highina, I. M. Bugaje, and B. Umar, "Biodiesel production from Jatropha caucus oil in a batch reactor using zinc oxide as catalyst,"

Journal of Petroleum Technology and Alternative Fuels, vol. 2, no. 9, pp. 146–149, 2011.

44. Supple, R. Howard-Hildige, E. Gonzalez-Gomez, and J. J. Leahy, "The effect of steam treating waste cooking oil on the yield of methyl ester," Journal of the American Oil Chemists' Society, vol. 79, no. 2, pp. 175–178, 2002.
45. Delfort, I. Durand, G. Hillion, A. Jaecker-Voirol, X. Montagne, Glycerin for new biodiesel formulation. Oil Gas Sci. Technol. e Rev. IFP 63(4), 395–404 (2008)
46. Dale, Greening the chemical industry: research and development priorities for biobased industrial products. J. Chem. Technol. Biotechnol. 78, 1093–1103 (2003)
47. B.B. He, A.P. Singh, J.C. Thompson, A novel continuous-flow reactor using reactive distillation for biodiesel production. Trans. ASAE 49, 107–112 (2006)
48. Bøhn T, Primicerio R, Hessen DO, Traavik T. Reduced fitness of Daphnia magna fed a bt-transgenic maize variety. Arch Environment Contamin Toxicol. 2008;55(4):584–592.
49. Boynton JE, Gillham NW. Genetics and transformation of mitochondria in the green alga Chlamydomonas. Methods Enzymol. 1996;264:279–296.
50. Boynton JE, Gillham NW. Chloroplast transformation in Chlamydomonas. Methods Enzymol. 1993;217:510–536.
51. Blanco AM, Moreno J, Del Campo JA, Rivas J, Guerrero MG. Outdoor cultivation of lutein-rich cells of Muriellopsis sp. in open ponds. Appl Microbiol Biotechnol. 2007;73(6):1259–1266.
52. Borevitz JO, Maloof JN, Lutes J, *et al.* Quantitative trait loci controlling light and hormone response in two accessions of Arabidopsis thaliana. Genetics. 2002;160(2):683–696.
53. Bingham SE, Cox JC, Strem MD. Expression of foreign DNA in Chlamydomonas reinhardtii. FEMS Microbiol Lett. 1989;53(1–2):77–81.
54. Bloomberg. Retrieved from http://www.bloomberg.com/news/2013-04-18/chevrondefies-california-on-carbon-emissions.html Fenton, O., & Ó hUallacháin, D. (2012). Agricultural nutrient

surpluses as potential input sources to grow third generation biomass (microalgae): A review. Algal Research, 1(1), 49-56. Georgianna, D. R., Hannon, M. J., Marcuschi, M., Wu, S., Botsch, K., Lewis, A. J., . . . Mayfield, S. P. (2013). Production of recombinant enzymes in the marine alga dunaliella tertiolecta. Algal Research, 2(1), 2-9. Handler, R. M., Canter, C. E., Kalnes, T. N., Lupton, F. S., Kholiqov, O., Shonnard, D. R., & Blowers, P. (2012).

55. Biotechnology Advances 25:294- 306. Chisti Y. 2008. Biodiesel from microalgae beats bioethanol. Trends in Biotechnology 26(3):126-131. Daigo K, Nakano Y, Casareto Be, Suzuki Y, Shioi Y. 2008.
56. Benthic cyanobacteria of the genus Nodularia are non-toxic, with gas vacuoles, able to glide and genetically more diverse than planktonic Nodularia.
57. Berry JP, Gantar M, Perez MH, Berry G, Noriega FG. Cyanobacterial toxins as allelochemicals with potential applications as algaecides, herbicides and insecticides. Mar Drugs. 2008;6:117–146.
58. Bhadury P, Wright PC. Exploitation of marine algae: biogenic compounds for potential antifouling applications. Planta. 2004;219(4):561–578.
59. Borowitzka MA. Culturing microalgae in outdoor ponds. In: Andersen RA, editor. Algal Culturing Techniques. Vol. 218. Academic Press; NY, USA: 2005.
60. Benthic cyanobacteria of the genus Nodularia are non-toxic, with gas vacuoles, able to glide and genetically more diverse than planktonic Nodularia. International Journal of Systematic Microbiology 55:555-568.
61. Bernstein L, Bosch P, Canziani 0, Chen Z, Christ R, Davidson O, Hare W, Huq S, Karoly D, Kattsov V, Kundzewicz Z, Liu J, Lohmann U, Manning M, Matsuno T, Menne B, Metz B, Mirza M, Nicholls N, Nurse L, Pachauri R, Palutikof J, Parry M, Qin D, Ravindranath N, Reisinger A, Ren J, Riahi K, Rosenzweig C, Rusticucci M, Schneider S, Sokona Y, Solomon S, Stott P, Stouffer R, Sugiyama T, Swart R, Tirpak D, Vogel C, Yohe G. 2007.
62. Barba FJ, Grimi N, Vorobiev E: New Approaches for the Use of Non-conventional Cell Disruption Technologies to Extract Potential

Food Additives and Nutraceuticals from Microalgae. Food Eng Rev.2015; 7(1):45–62. 10.1007/s12393-014-9095-6

63. Borowitzka MA: High-value products from microalgae—their development and commercialisation. J Appl Phycol. 2013; 25(3):743–56. 10.1007/s10811-013-9983-9
64. Bhatnagar A, Chinnasamy S, Singha M, Das KC. Renewable biomass production by mixotrophic algae in the presence of various carbon sources and wastewaters. Appl Energy in press. doi:10.1016/j.apenergy.2010.12.064
65. Borowitzka MA. Algal biotechnology products and processes – matching science and economics. J Applied Phycology. 1992; 4:267–279.
66. Borowitzka MA. Commercial production of microalgae: ponds, tanks, tubes and fermenters. J Biotechnology. 1999; 70:313–321. Krichnavaruk S, Shotipruk A, Goto M, Pavasant P.
67. Byerlee D, De Janvry A, World Bank. Agriculture for Development. World Bank; DC, USA: 2007.
68. Berman-Frank I, Lundgren P, Falkowski P. Nitrogen fixation and photosynthetic oxygen evolution in cyanobacteria. Res Microbiol. 2003; 154(3):157–164.
69. Boyd PW, Law CS, Wong CS, *et al.* The decline and fate of an iron-induced subarctic phytoplankton bloom. Nature. 2004; 428(6982):549–553.
70. Brussaard CPD. Viral control of phytoplankton populations – a review. J Eurkaryotic Microbiol. 2004; 51(2):125–138.
71. Baumann, H., andA.M. Tillman. 2014. The Hitch Hikers Guide to LCA. Lund, Sweden: Studentliterature AB.
72. Benemann, J.R., R.P. Goebel, J.c Weissman, and D.C Augenstein.1982. Microalgae as a source of Liquid Fuel. FairField, CA: EnBio, Inc.
73. Bhatnagar, A., M. Bhapnagar, S.Chinnasamy, and K.C. Das. 2010. Chlorella minutissima- A promosing fuel alga for cultivation in municipal wastewater. Applied biochemistry and biotechnology 161(1-8):523.536.

74. Brune, D.E., T.J. Lundquist, and J.R. Benemann. 2009. Microalgal biomass for greenhouse gas reductions:

75. Bullard, C.W., and R.A.Herendeen. 1975. Energy cost of goods and services. Energy policy 3(4):268-278.

76. Bullard, C., P.Pennter, and D, Pilati. 1978. Net energy analysis: Handbook for combining process and input-output analysis Resources and Energy 1:267-313.

77. Boussiba, S. 2011. Questionnaire reply from AlgaTechnologies, Kibbutz Qetura. Received by the NRC Committee on the Sustainable Development of Algal Biofuels on July 6.

78. Beal, C.M., R.E. Hebner, M.E. Webber, R.S. Ruoff, and A.F. Seibert. 2011. The energy return on investment for algal biocrude: Results for a research production facility. Bioenergy Research: DOI: 10.1007/s12155-011-9128-4.

79. Benemann, J.R. 2008. Opportunities and challenges in algae biofuel production. Available online at http://www.fao.org/uploads/media/algae_positionpaper.pdf. Accessed October 21, 2011.

80. BioProcess Algae. 2011. Questionnaire reply from BioProcess Algae. Received by the NRC Committee on the Sustainable Development of Algal Biofuels on September 27.

81. Baral, A., B.R. Bakshi, and R.L. Smith. 2012. Assessing resource intensity and renewability of cellulosic ethanol technologies using Eco-LCA. Environmental Science and Technology 46 (4):2436-2444.

82. Batan, L., J. Quinn, T. Bradley, and B. Willson. 2011. Erratum: Net energy and greenhouse gas emissions evaluation of biodiesel derived from microalgae (Environmental Science and Technology (2010) 44 (7975-7980)). Environmental Science and Technology 45(3):1160.

83. Batan, L., J. Quinn, B. Willson, and T. Bradley. 2010. Net energy and greenhouse gas emission evaluation of biodiesel derived from microalgae. Environmental Science and Technology 44(20):7975-7980.

84. Benemann, J., P.M. Pedroni, J. J. Davison, H. Beckert, and P. Bergman. 2003a. Technology roadmap for biofixation of CO2 and

greenhouse gas abatement with microalgae. Paper read at the Second Annual Conference on Carbon Sequestration, May 5-8, Alexandria, VA.

85. Benemann, J.R., J.C. Van Olst, M.J. Massingill, J.C. Weissman, and D.E. Brune. 2003b. The controlled eutrophication process: Using microalgae for CO2 utilization and agricultural fertilizer recycling. Paper read at the 6th Greenhouse Gas Control Technologies Conference (GHGT6), October, Kyoto Japan.
86. BLM (Bureau of Land Management) and DOE (U.S. Department of Energy). 2010. Solar Energy Development Draft Programmatic Environmental Impact Statement (Draft Solar PEIS). Washington, D.C.: Bureau of Land Management and U.S. Department of Energy.
87. Boyd, C.E., and A. Gross. 2000. Water use and conservation for inland aquaculture ponds. Fisheries Management and Ecology 7(1-2):55-63.
88. Brentner, L.B., M.J. Eckelman, and J.B. Zimmerman. 2011. Combinatorial life cycle assessment to inform process design of industrial production of algal biodiesel. Environmental Science and Technology 45:7060-7067.
89. Babcock, B.A. 2009. Measuing unmeasurable land-use changes from biofuels. Iowa Agriculture Review Summer: 4-6.
90. Bachmann, R.W., B.L. Jones, D.D. Fox, M. Hoyer, L.A. Bull, and D.E. Canfield. 1996. Relations between trophic state indicators and fish in Florida (USA) lakes. Canadian Journal of Fisheries and Aquatic Sciences 53(4):842-855.
91. Beeman, P. 2007. Biofuel plants generate new air, water, soil problems for Iowa. Available onlineat http://www.desmoinesregister.com/article/20070603/BUSINESS01/706030325/Biofuel-plants-generate-new-air-water-soil-problems-for-Iowa. Accessed January 21, 2012.
92. Benemann, J.R. 2008. Opportunities and challenges in algae biofuel production. Available online at http://www.fao.org/uploads/media/algae_positionpaper.pdf. Accessed October 21, 2011.
93. BLM (Bureau of Land Management) and DOE (U.S. Department of Energy). 2010. Solar Energy Development Draft Programmatic

Environmental Impact Statement (Draft Solar PEIS). Washington, DC: Bureau of Land Management and U.S. Department of Energy.

94. Bollag, J.M., and S.T. Czlonkowski. 1973. Inhibition of methane formation in soil by various nitrogen-containing compounds. Soil Biology and Biochemistry 5(5):673-678.
95. Boyd, C.E. 1995. Bottom soils, sediment, and pond aquaculture. New York: Chapman and Hall.
96. Breitburg, D.L., D.W. Hondorp, L.A. Davias, and R.J. Diaz. 2009. Hypoxia, nitrogen, and fisheries: Integrating effects across local and global landscapes. Pp. 329-349 in Annual Review of Marine Science.
97. Briand, E., C. Yepremian, J.F. Humbert, and C. Quiblier. 2008. Competition between microcystin- and non-microcystin-producing Planktothrix agardhii (cyanobacteria) strains under different environmental conditions. Environmental Microbiology 10(12):3337-3348.
98. Buntjer, J. 2010. Heron Lake BioEnergy fined. Daily Globe, December 27. State and Regional News section.
99. Burkholder, J.M., K.M. Mason, and H.B. Glasgow. 1992. Water-column nitrate enrichment promotes decline of eelgrass Zostera-Marina—Evidence from seasonal mesocosm experiments. Marine Ecology-Progress Series 81(2):163-178.
100. Burns, J.C. 2008. ASAS centennial paper: Utilization of pasture and forages by ruminants: A historical perspective. Journal of Animal Science 86(12):3647-3663.
101. Benthic cyanobacteria of the genus Nodularia are non-toxic, with gas vacuoles, able to glide and genetically more diverse than planktonic Nodularia.
102. Biotechnology Advances 25:294- 306. Chisti Y. 2008. Biodiesel from microalgae beats bioethanol. Trends in Biotechnology 26(3):126-131. Daigo K, Nakano Y, Casareto Be, Suzuki Y, Shioi Y. 2008.
103. Biodiesel Production Technology. USA: NREL (SR-510-36244) http://www.nrel.gov/docs/fy04osti/36244.pdf (accessed on 10th December 2012) Gouveia L, 2011.

104. Bernstein L, Bosch P, Canziani 0, Chen Z, Christ R, Davidson O, Hare W, Huq S, Karoly D, Kattsov V, Kundzewicz Z, Liu J, Lohmann U, Manning M, Matsuno T, Menne B, Metz B, Mirza M, Nicholls N, Nurse L, Pachauri R, Palutikof J, Parry M, Qin D, Ravindranath N, Reisinger A, Ren J, Riahi K, Rosenzweig C, Rusticucci M, Schneider S, Sokona Y, Solomon S, Stott P, Stouffer R, Sugiyama T, Swart R, Tirpak D, Vogel C, Yohe G. 2007. Climate Change 2007: Synthesis Report – Summary for Policymakers.

105. Biotechnology Advances, 30(3), 673-690. Lane, J. (2013, April 9). A crash of algae, bioterrorism and the future of biofuels. Biofuels Digest. Retrieved from http://www.biofuelsdigest.com/bdigest/2013/04/09/a-crashof-algae-bioterrorism-and-the-future-of-biofuels/ Mahapatra, D. M., Chanakya, H. N., & Ramachandra, T. V. (2013).

106. Barros AI, Gonçalves AL, Simões M, *et al.* : Harvesting techniques applied to microalgae: A review.Renewable and Sustainable Energy Reviews. 2015;41:1489–500. 10.1016/j.rser.2014.09.037

107. Burkart MD: Green algae cultivation and engineering of its fatty acid synthase. In: 249th American Chemical Society National Meeting & Exposition.2015.

108. Barrick JE, Yu DS, Yoon SH, *et al.* : Genome evolution and adaptation in a long-term experiment withEscherichia coli. Nature. 2009;461(7268):1243–7. 10.1038/nature08480

109. Beyter D, Tang PZ, Becker S, *et al.* : Diversity, Productivity, and Stability of an Industrial Microbial Ecosystem. Appl Environ Microbiol. 2016;82(8):2494–505. 10.1128/AEM.03965-15

110. Biodiesel from microalgae.

111. Biotechnology Advances 25, 294–306 9. Huber, G.W., Iborra, S., Corma, A.(2006).

112. Coale KH, Johnson KS, Chavez FP, *et al.* Southern ocean iron enrichment experiment: carbon cycling in high- and low-si waters. Science. 2004; 304(5669):408–414.

113. Riordon, G. Love, N. Barron, P. Nigam, R. Marchant, L. McHale *et al.*, Production of ethanol from sucrose at 45 °C by alginate immobilized preparations of the thermotolerant yeast strain Kluyveromyces marxianus IMB 3. Bioresour. Technol. 55, 17–173

(1996)

114. C.L. Peterson, J.L. Cook, J.C. Thompson, J.S. Taberski, Continuous flow biodiesel production. Appl. Eng. Agric. 18(1), 5–11 (2002)

115. Coale KH, Johnson KS, Fitzwater SE, *et al.* A massive phytoplankton bloom induced by an ecosystem-scale iron fertilization experiment in the equatorial pacific ocean. Nature. 1996; 383(6600):495–501.

116. Chermack, T.J., S.A. Lynham and W.E.A. Ruona. A review of scenario planning literature. Available online at http://www.cse.buffalo.edu/-peter/refs/DataAssimilation/Multihypoyhesis/ReviewofSP.PDF. Accessed June 18, 2012.

117. Chinnasamy, S., A. Bhatnagar, R.W. Hunt, and K.C. Das.2010. Microalgae cultivation in wastewater dominated by carpet mill effluents for biofuel applications. Bioresource Technology 101(9):3097-3105.

118. Corbiere-Nicollier, T., I, Blanc, and S. Erkman. 2011. Towards a global criteria based framework for the sustainability assessment of bioethanol supply chains. Application to the swiss dilemma: Is local produced bioethanolmore sustainable than bioethanol imported from Brazil? Ecological indicators 11:1447-1458.

119. Craggs, R, I., S. Heubeck, T.J. Lundquist, and J.R. Benemann. 2011. Algal biofuels from wastewater treatment high rate algal ponds. Water Science and Technology 63(4):660-665.

120. CAI, X.M., X.A. Zhang, and D.B. Wang. 2011. Land availability for biofuel production. Environmental Science and Technology 45(1):334-339.

121. Campbell, P.K., T. Beer, and D. Batten. 2011. Life cycle assessment of biodiesel production from microalgae in ponds. Bioresource Technology 102(1):50-56.

122. Chiu, Y.-W., B. Walseth, and S. Suh. 2009. Water embodied in bioethanol in the United States. Environmental Science and Technology 43(8):2688-2692.

123. Clarens, A.F., H. Nassau, E.P. Resurreccion, M.A. White, and L.M. Colosi. 2011. Environmental impacts of algae-derived biodiesel and bioelectricity for transportation. Environmental Science and Technology 45:7554-7560.

124. Clarens, A.F., E.P. Resurreccion, M.A. White, and L.M. Colosi. 2010. Environmental life cycle comparison of algae to other bioenergy feedstocks. Environmental Science and Technology 44(5):1813-1819.

125. Cordell, D., J.O. Drangert, and S. White. 2009. The story of phosphorus: Global food security and food for thought. Global Environmental Change-Human and Policy Dimensions 19(2):292-305.

126. Craggs, R.J., S. Heubeck, T.J. Lundquist, and J.R. Benemann. 2011. Algal biofuels from wastewater treatment high rate algal ponds. Water Science and Technology 63(4):660-665.

127. Chisti, Y. 2007. Biodiesel from microalgae. Biotechnology Advances 25:294-306.

128. Chance, R.R., B. McCool, and J.D. Coleman. 2011. A Cyanobacteria-Based Photosynthetic Process for the Production of Ethanol. Presentation to the NRC Committee on Sustainable Development of Algal Biofuels on June 13.

129. Carpenter EJ, Janson S. Intracellular cyanobacterial symbionts in the marine diatom Climacodium frauenfeldianum (bacillariophyceae) J Phycol. 2000;36(3):540–544.

130. Chaisson Mj, Brinza D, Pevzner PA. De novo fragment assembly with short mate-paired reads: does the read length matter? Genome Res. 2009;19(2):336–346.

131. Casas-Mollano JA, Van Dijk K, Eisenhart J, Cerutti H. SET3P monomethylates histone H3 on lysine 9 and is required for the silencing of tandemly repeated transgenes in Chlamydomonas. Nucleic Acids Res. 2007;35(3):939–950.

132. Charles, C. & Wooders, P. (2012). Biofuels – At what cost? Mandating ethanol and biodiesel consumption in the United Kingdom. Geneva: International Institute for Sustainable Development (IISD). Chowdhury, R., Viamajala, S., & Gerlach, R. (2012). Reduction of environmental and energy footprint of microalgal biodiesel production through material and energy integration. Bioresource Technology, 108, 102-111.

133. Chisti Y. Biodiesel from microalgae beats bioethanol. Trends Biotechnol. 2008;26(3):126–131

134. Chisti Y. Fuels from microalgae. Biofuels 2010;1:233–5.

135. Climate Change 2007: Synthesis Report – Summary for Policymakers. Intergovernmental Panel on Climate Change. http://www.ipcc.ch/pdf/assessment-report/ar4/syr/ar4_syr.pdf (accessed on 12th October 2013). Campbell A, Cooke P, Cass K, Earl K. 2004. Plant and Animal Evolution. New Zealand, the University of Waikato. http://sci.waikato.ac.nz/evolution/aboutus.shtml (accessed on 2nd April 2013) Charpy L, Palinska Ka, Casareto B, Langlade Mj, Suzuki Y, Abed Rmm, Golubic S. 2010. Dinitrogen-Fixing Cyanobacteria in Microbial Mats of Two Shallow Coral Reef Ecosystems. Microb Ecol 59:174-186.

136. Conley DJ, Paerl HW, Howarth RW, *et al.* Ecology. Controlling eutrophication: nitrogen and phosphorus. Science. 2009;323(5917):1014–1015.

137. Chisti Y. Biodiesel from microalgae. Biotechnol Adv 2007;25:294–306.

138. Chisti Y. Biodiesel from microalgae beats bioethanol. Trends Biotechnol 2008;26:126–31.

139. Chen F, Li H, Wong R N, Jiang Y. 2005. Isolation and purification of the bioactive carotenoid zeaxanthin from the microalga Microcystis aeruginosa by high-speed counter-current chromatography.

140. Chu CY, Liao WR, Huang R, Lin LP. Haemagglutinating and antibiotic activities of freshwater microalgae. World J Microbiol Biotechnol. 2004;20(8):817–825.

141. Chen Y, Wang Y, Sun Y, Zhang L, Li W. Highly efficient expression of rabbit neutrophil peptide-1 gene in chlorella ellipsoidea cells. Curr Genet. 2001;39(5):365–370.

142. Clayton, M., "New Coal Plants Bury 'Kyoto,'" Christian Science Monitor, December 23, 2004. Christi, Y., "Biodiesel from Microalgae," Biotechnology Advances, Vol. 25, 2007, pp. 294-306. Crookes, R. J., "Comparative Bio-Fuel Performance in Internal Combustion Engines," Biomass and Bioenergy, Vol. 30, 2006, pp. 461-468.

143. Copping LG, Duke SO. Natural products that have been used commercially as crop protection agents. Pest Management Sci.

2007;63(6):524–554.

144. Chisti Y. Biodiesel from microalgae. Biotechnol Adv. 2007;25(3):294–306.

145. Chisti Y. Biodiesel from microalgae. Biotechnol Adv 2007;25:294–306.

146. Chisti Y. Biodiesel from microalgae beats bioethanol. Trends Biotechnol 2008;26:126–31.

147. Chisti Y. Biodiesel from microalgae. Biotechnol Adv. 2007;25(3):294–306.

148. Degray G, Rajasekaran K, Smith F, Sanford J, Daniell H. Expression of an antimicrobial peptide via the chloroplast genome to control phytopathogenic bacteria and fungi. Plant Physiol. 2001;127(3):852.

149. Dunahay TG, Jarvis EE, Roessler PG. Genetic transformation of the diatoms Cyclotella cryptica and Navicula saprophila. J Phycol. 1995;31(6):1004–1012.

150. De Riso V, Raniello R, Maumus F, Rogato A, Bowler C, Falciatore A. Gene silencing in the marine diatom Phaeodactylum tricornutum. Nucleic Acids Res. 2009;37(14):e96.

151. Department of Energy's Office of Fuels Development NREL/TP-580-24190, July 1978. U.S. Department of Energy, "Alternative Fuels & Advanced Vehicles Data Center", Last updated: Nov 12, 2007, tonto.eia.doe.gov/oog/info/gdu/gasdiesel.asp, Accessed: Nov 14, 2007. Xu, H.; Miao, X. and Q. Wu., "High Quality Biodiesel Production from a Microalga Chlorella protothecoides by Heterotrophic Growth in Fermenters," Journal of Biotechnology, Vol. 126, 2006, pp.499-507.

152. Y. C. Leung, X. Wu, and M. K. H. Leung, "A review on biodiesel production using catalyzed transesterification," Applied Energy, vol. 87, no. 4, pp. 1083–1095, 2010.

153. Darnoko, M. Cheryan, Continuous production of palm methyl esters. J. Am. Oil Chem. Soc. 77, 1269–1272 (2000)

154. Dragone G. Fernandes BD. Abreu AP. Vicente AA. Teixeira JA. Nutrient limitation as a strategy for increasing starch accumulation in microalgae. Appl Energy in press. doi:10.1016/j.apenergy.2011.03.012.

155. Dyni JR. Scientific Investigations Report 2005-5294. US Geological Survey; VA, USA: 2006. Geology and resources of some world oil-shale deposits.

156. Deschamps P, Moreira D. Signal conflicts in the phylogeny of the primary photosynthetic eukaryotes. Mol Biol Evol. 2009;26(12):2745–2753.

157. Douskova I, Doucha J, Livansky K, *et al.* Simultaneous flue gas bioremediation and reduction of microalgal biomass production costs. Appl Microbiol Biotechnol. 2009;82(1):179–185.

158. Murray, Ethanol's potential: looking beyond corn (Earth Policy Institute, Washington DC, 2005). http://www.earthpolicy.org/Updates/2005/Update49.htm

159. D. Brady, P. Nigam, R. Marchant, A.P. McHale, Ethanol production at 45 °C by immobilized Kluyveromyces marxianus IMB3 during growth on lactose-containing media. Bioprocess Eng. 16, 101–104 (1997)

160. D. Brady, P. Nigam, R. Marchant, D. Singh, A.P. McHale, The effect of Mn2þ on ethanol production from actose using Kluyveromyces marxianus IMB3 immobilized in magnetically responsive matrices. Bioprocess Eng. 17, 31–34 (1997)

161. D. Kusdiana, S. Saka, Kinetics of transesterification in rapeseed oil to biodiesel fuels as treated in supercritical methanol. Fuel 80, 693–698 (2001)

162. D. Kusdiana, S. Saka, Two-step preparation for catalyst-free biodiesel fuel production: hydrolysis and methyl esterification. Appl. Biochem. Biotechnol. 115, 781–792 (2004)

163. D. Royon, M. Daz, G. Ellenrieder, S. Locatelli, Enzymatic production of biodiesel from cotton seed oil using t-butanol as a solvent. Bioresour. Technol. 98, 648–653 (2007)

164. D. Brady, P. Nigam, R. Marchant, L. McHale, A.P. McHale, Ethanol production at 45 °C by Kluyveromyces marxianus IMB3 immobilized in magnetically responsive alginate matrices. Biotechnol. Lett. 18(10), 1213–1216 (1996)

165. D. Bajpai, V.K. Tyagi, Biodiesel: source, production, composition, properties and its benefits. J. Olio Sci. 55, 487–502 (2006)

166. Dubinsky Z, Rotem J. Relations between algal populations and the pH of their media. Oecologia. 1974;16:53–69.

167. Dalrymple, O. K., Halfhide, T., Udom, I., Gilles, B., Wolan, J., Zhang, Q., & Ergas, S. (2013). Wastewater use in algae production for generation of renewable resources: A review and preliminary results. Aquatic Biosystems, 9(1), 1-11. Dana, G. V., Kuiken, T., Rejeski, D., & Snow, A. A. (2012). Synthetic biology: Four steps to avoid a synthetic-biology disaster. Nature, 483(7387), 29. Elgin, B. & Waldman, P. (2013, April 18). Chevron Defies California On Carbon Emissions.

168. Debuchy R, Purton S, Rochaix JD. The argininosuccinate lyase gene of Chlamydomonas reinhardtii: an important tool for nuclear transformation and for correlating the genetic and molecular maps of the Arg7 locus. EMBO J. 1989;8(10):2803–2809.

169. Dinitrogen-Fixing Cyanobacteria in Microbial Mats of Two Shallow Coral Reef Ecosystems. Microb Ecol 59:174-186. Chen F, Li H, Wong R N, Jiang Y. 2005.

170. Davis, R., A. Aden, and P.T. Pienkos. 2011. Techno-economic analysis of autotrophic microalgae for fuel production. Applied Energy 88(10):3524-3531.

171. Demirbas, A., "Importance of Biodiesel as Transportation Fuel," Energy Policy, Vol. 35, 2007, pp. 4661-4670.

172. Dismukes GC, Carrieri D, Bennette N, Ananyev GM, Posewitz MC. Aquatic phototrophs: efficient alternatives to land-based crops for biofuels. Curr Opin Biotechnol. 2008;19(3):235–240.

173. Development in oil extraction from microalgae.

174. Demirbas MF. Biofuels from algae for sustainable development. Appl Energy, in press. doi:10.1016/j.apenergy.2011.01.059.

175. Dismukes GC, Carrieri D, Bennette N, Ananyev GM, Posewitz MC. Aquatic phototrophs: efficient alternatives to land-based crops for biofuels. Curr Opin Biotechnol. 2008;19(3):235–240.

176. Doucha J, Livansky K. Productivity, CO2/O2 exchange and hydraulics in outdoor open high density microalgal (Chlorella sp.) photobioreactors operated in a middle and southern European climate. J Applied Phycology. 2006;18(6):811–826.

177. Energy Information Administration. International Energy Outlook. Vol. 284. EIA; DC, USA: 2009.

178. European Journal of Lipid Science and Technology 113(5):539-547. Nakpong P, Wootthikanokkhan S. 2010.

179. Engineering Journal 16(5):158-166.

180. Ehimen EA, Sun ZF, Carrington CG, Birch EJ, Eaton-Rye J. Anaerobic digestion of microalgae residues resulting from the biodiesel production process. Appl Energy in press. doi:10.1016/j.apenergy.2010.10.020.

181. Energy return on investment for algal biofuel production coupled with wastewater treatment. Water Environment Research, 84(9), 692-710. Borowitzka, M. A., & Moheimani, N. R. (2013). Sustainable biofuels from algae. Mitigation and Adaptation Strategies for Global Change, 18(1), 13-25. Cai, T., Park, S. Y., & Li, Y. (2013). Nutrient recovery from wastewater streams by microalgae: Status and prospects. Renewable and Sustainable Energy Reviews, 19, 360- 369. Casey, T. (2013, February 8). EU Is Burning More Coal, Looks To "Burn" Seawater Instead. CleanTechnica. Algal biofuel production and mitigation potential in India. Mitigation and Adaptation Strategies for Global Change, 18(1), 113-136.

182. E.D. Larson, Biofuel production technologies: status, prospects and implications for trade and development. Report No. UNCTAD/DITC/TED/2007/10. United Nations Conference on Trade and Development, New York and Geneva, 2008

183. Lotero, Y.J. Liu, D.E. Lopez, K. Suwannakarn, D.A. Bruce, J.G. Goodwin, Synthesis of biodiesel via acid catalysis. Ind. Eng. Chem. Res. 44, 5353–5363 (2005)

184. Minami, S. Saka, Kinetics of hydrolysis and methyl esterification for biodiesel production in two-step supercritical methanol process. Fuel 85, 2479–2483 (2006)

185. EU, Directive 2009/28/EC of The European Parliament and of The Council of 23 April 2009 on the promotion of the use of energy from renewable sources and amending and subsequently repealing Directives 2001/77/EC and 2003/30/EC. Off. J. Eur. Union 52, 16–62 (2009)

186. Evaluation of environmental impacts from microalgae cultivation in open-air raceway ponds: Analysis of the prior literature and investigation of wide variance in predicted impacts. Algal Research, 1(1), 83-92. Henley, W. J., Litaker, R. W., Novoveská, L., Duke, C. S., Quemada, H. D., & Sayre, R. T. (2012).

187. Euglena sp. as a suitable source of lipids for potential use as biofuel and sustainable wastewater treatment. Journal of Applied Phycology, January 27, 1-11.

188. Energy Information Administrator, "Gasoline and Diesel Fuel Update", Last updated: Oct 1, 2007, www.eere.energy.gov/afdc/price_report.html, Accessed: Nov 14, 2007. Haag, A. L., "Algae Bloom Again," Nature, Vol. 447, May 31, 2007, pp. 520-521.

189. Energy Information Administration. International Energy Outlook. Vol. 284. EIA; DC, USA: 2009.

190. Fan, X., R. Burton, and G. Austic. 2010. Conversion of degummed soybean oil to biodiesel: Optimization of degumming methods and evaluation of fuel properties. International Journal of Green Energy 7(6):593-599.

191. Flynn KJ, Mitra A, Greenwell HC, *et al.* : Monster potential meets potential monster: pros and cons of deploying genetically modified microalgae for biofuels production. Interface Focus. 2013;3(1):20120037. 10.1098/rsfs.2012.0037

192. Fuel 89:1806-1811. Noaa Earth System Research Laboratory. 2013.

193. Fargione J, Hill J, Tilman D, Polasky S, Hawthorne P. Land clearing and the biofuel carbon debt. Science. 2008;319(5867):1235–1238.

194. Falkowski PG, Barber RT, Smetacek VV. Biogeochemical controls and feedbacks on ocean primary production. Science. 1998;281(5374):200–207.

195. Fierro S, Sanchez-Saavedra Mdel P, Copalcua C. Nitrate and phosphate removal by chitosan immobilized Scenedesmus. Bioresour Technol. 2008;99(5):1274–1279.

196. Ullah, A. Nosheen, I. Hussain, and A. Bano, "Base catalyzed transesterification of wild apricot kernel oil for biodiesel production," African Journal of Biotechnology, vol. 8, no. 14, pp. 3289–3293, 2009.

197. O. Licht, "The Global Renewable Fuels Alliance is a non-profit organization dedicated to promoting biofuel friendly policies internationally," Global Renewable Fuels Alliance, 2011.

198. Funderberg E. Why are nitrogen prices so high. Pasture Range. 2009;5

199. F.O. Licht, World ethanol & biofuels report (Agra Informa Ltd., Kent). http://www.agra-net.com/portal/puboptions.jsp?Option¼menu&pubId¼ag072

200. FAO, The state of food and agriculture. BIOFUELS: prospects, risks and opportunities(FAO, Rome, 2008)

201. F. Omota, A.C. Dimian, A. Bliek, Fatty acid esterification by reactive distillation. Part 1: equilibrium-based design. Chem. Eng. Sci. 58, 3159–3174 (2003)

202. FAO, in World agriculture: towards 2015/2030. An FAO perspective, ed. by J. Bruinsma. Food and Agriculture Organization (FAO) (Earthscan Publications Ltd, London, 2003).

INDEX